The "Resource Curse" in the Persian Gulf

T0229524

The "Resource Curse" in the Persian Gulf systematically addresses the little studied notion of a "resource curse" in relation to the Persian Gulf by examining the historical causes and genesis of the phenomenon and its consequences in a variety of areas, including human development, infrastructural growth, clientelism, state-building and institutional evolution, and societal and gender relations.

The book explores how across the Arabian Peninsula, oil wealth began accruing to the state at a particular juncture in the state-building process, when traditional, largely informal patterns of shaikhly rule were relatively well established, but the formal institutional apparatuses of the state were not yet fully formed. The chapters show that oil wealth had a direct impact on subsequent developments in these two complementary areas. Contributors discuss how on one hand, the distribution of petrodollars enabled political elites to solidify existing patterns of rule through deepening clientelist practices and by establishing new, dependent clients; and how on the other, rent revenues gave state leaders the opportunity to establish and shape institutions in ways that solidified their political control.

The "Resource Curse" in the Persian Gulf will be of great interest to scholars of Middle Eastern studies, focusing on a variety of subject areas, including human development, human resources, clientelism, infrastructural growth, institutional evolution, state-building, and societal and gender relations. This book was originally published as a special issue of the *Journal of Arabian Studies*.

Mehran Kamrava is Professor and Director of the Center for International and Regional Studies (CIRS) at Georgetown University, Qatar. He is author of a number of books, including, most recently, *A Concise History of Revolution* (2020), *Inside the Arab State* (2018) and *Troubled Waters: Insecurity in the Persian Gulf* (2018).

The "Resource Curse" in the Persian Gulf

Edited by
Mehran Kamrava

 Routledge
Taylor & Francis Group

LONDON AND NEW YORK

جامعة جورجتاون قطر
GEORGETOWN UNIVERSITY QATAR
Center *for* International *and* Regional Studies

First published 2020
by Routledge
2 Park Square, Milton Park, Abingdon, Oxon, OX14 4RN

and by Routledge
52 Vanderbilt Avenue, New York, NY 10017

Routledge is an imprint of the Taylor & Francis Group, an informa business

First issued in paperback 2021

British Library Cataloguing in Publication Data
A catalogue record for this book is available from the British Library

ISBN13: 978-0-367-40874-9 (hbk)
ISBN13: 978-1-03-208845-7 (pbk)

Typeset in Times
by Newgen Publishing UK

Publisher's Note
The publisher accepts responsibility for any inconsistencies that may have arisen during
the conversion of this book from journal articles to book chapters, namely the inclusion
of journal terminology.

Disclaimer
Every effort has been made to contact copyright holders for their permission to reprint
material in this book. The publishers would be grateful to hear from any copyright holder
who is not here acknowledged and will undertake to rectify any errors or omissions in
future editions of this book.

Contents

Citation Information

The chapters in this book were originally published in the *Journal of Arabian Studies*, volume 8, supplement 1 (September 2018). When citing this material, please use the original page numbering for each article, as follows:

Chapter 6
The Impact of Oil Rents on Military Spending in the GCC Region: Does Corruption Matter?
Mohammad Reza Farzanegan
Journal of Arabian Studies, volume 8, supplement 1 (September 2018) pp. 87–109

For any permission-related enquiries please visit:
www.tandfonline.com/page/help/permissions

Notes on Contributors

Gail J. Buttorff is Assistant Professor in the Department of Political Science at the University of Kansas, USA.

Mohammad Reza Farzanegan is Professor of the Economics of Middle East at the interdisciplinary Center for Near and Middle Eastern Studies (CNMS) at the Marburg Centre for Institutional Economics, Germany.

Desha Girod is Associate Professor and Director of the M.A. in Conflict Resolution in the Department of Government at Georgetown University, USA.

Matthew Gray is Associate Professor in the School of International Liberal Studies at Waseda University, Tokyo.

Mehran Kamrava is Professor and Director of the Center for International and Regional Studies (CIRS) at Georgetown University, Qatar.

Nawra Al Lawati is a post graduate research student at the University of Liverpool, UK.

Jessie Moritz is a lecturer at the Centre for Arab and Islamic Studies at the Australian National University, Australia.

Meir R. Walters is a Foreign Affairs Officer in the Bureau of Democracy, Human Rights, and Labor's (DRL) Office of Near Eastern Affair at the U.S. Department of State, USA.

Bozena C. Welborne is Assistant Professor of Government at Smith College, USA.

1 Oil and Institutional Stasis in the Persian Gulf

Mehran Kamrava

Abstract: Oil has seriously impacted the institutional development of the state in the Arabian Peninsula. More specifically, the sudden and unprecedented acquisition of massive oil revenues resulted in the freezing of the state's formal and informal institutions, at the point at which petrodollars were injected into the state's coffers. From then on, state leaders were able to deploy the state's wealth to dictate the pace and direction of institutional change. Over time, any institutional change has been directed towards enhancing regime security, and the pace of change has been calculated and deliberately slow. Any political opening has been dictated by the logic of state power maximization (in relation to society). At the same time, partly to ensure its popular legitimacy and partly through the vision of its leaders, the state has deployed its massive wealth both to foster rapid economic and infrastructural development, and to enhance the living standards of its citizens. In other words, whereas oil may have stunted institutional development — i.e., an institution's curse — it has been an economic blessing.

1 Introduction

The debate on whether resource abundance in general, and resource dependence in particular, is a curse or a blessing is an old one. Paradoxically, despite a proliferation of studies on rentierism in the last two decades or so,[1] the specific notion of a "resource curse" has seldom been studied systematically in relation to the Persian Gulf. The articles in this special issue address this gap, looking specifically at the historical causes and genesis of the phenomenon and its consequences in a variety of areas, including human development, infrastructural growth, clientelism, state-building and institutional evolution, and societal and gender relations. In this article, I introduce some of the broader aspects of the resource curse phenomenon in relation to the evolution of states in the Persian Gulf region. More specifically, I argue that one of the most pronounced

---

Author's note: This special issue is the result of a two-year research initiative undertaken by the Center for International and Regional Studies (CIRS) at Georgetown University in Qatar. The project was developed and supported by the CIRS team, all of whom deserve my heartfelt thanks: Zahra Babar, Suzi Mirgani, Elizabeth Wanucha, Jackie Starbird, Islam Hassan, and Misba Bhatti. I also acknowledge the participation of my colleagues Gerd Nonneman and Anatol Lieven, who enriched the working group discussions. Finally, grateful acknowledgment also goes to the Qatar Foundation for its support of research and other scholarly endeavors.

[1] Elbadawi and Soto, "Resource Rents, Political Institutions and Economic Growth", *Understanding and Avoiding the Oil Curse in Resource-Rich Arab Economies*, ed. Elbadawi and Selim (2016), p. 188.

consequences of resource abundance and dependence in the region has been a freezing of state institutions — especially those institutions through which political power is exercised — at the moment in which oil revenues began flowing into the coffers of the state.

Across the Arabian Peninsula, oil wealth began accruing to the state at a particular juncture in the state-building process, when traditional, largely informal patterns of shaikhly rule were relatively well established but the formal institutional apparatuses of the state were not yet fully formed. Oil wealth had a direct impact on subsequent developments in these two complementary areas. On the one hand, the control and distribution of petrodollars enabled political elites to solidify existing patterns of rule through deepening clientelist practices and also by establishing new, dependent clients. On the other hand, rent revenues gave state leaders the opportunity to establish and shape state institutions in ways that solidified their political control. Tribally rooted patron-client practices were superimposed upon and reinforced through an ostensibly modern bureaucratic apparatus. Rent-seeking, clientelism, and patronage dictated the logic of institutional change and evolution.

A combination of rent-seeking and patron-client relations froze some institutions in place, slowed or skewed the change that other institutions might otherwise have experienced, and enabled state leaders to shape the direction and pace at which institutional change occurred. The consequence was neopatrimonialism, which informed, and continues to inform, the institutional makeup and functions of the state and its larger approach towards society. If one adopts the terminology commonly used to refer to the consequences of resource dependence, whatever "curse" has befallen the states of the Arabian Peninsula has been more institutional than economic or developmental.

Before elaborating on these points, a few preliminary remarks about the overall causes and consequences of the resource curse are in order. The article then focuses on the specific relationship between resource abundance and the development of state institutions in the Persian Gulf region. In addition to providing a summary of the analysis here and in the other articles in this issue, the conclusion will offer some thoughts on future directions for research on the topic.

2 Whither the resource curse?

An important early observation is that the resource curse is not inevitable and is, in fact, conditional on bad governance.[2] Political institutions in general and fiscal institutions in particular can help turn a resource curse into a blessing.[3] By itself, wealth does not generate lack of financial discipline or pathologies such as corruption or inefficiency. For example, one of the negative side effects of reliance on resources is living off capital rather than income, and relying on revenues from a nonrenewable source. One remedy would be to turn those earnings into financial assets, investing those assets into a diversified portfolio, and treating the interests accrued as income.[4] As Elbadawi and Selim have argued, "while macroeconomic mismanagement and oil abundance are important determinants of performance, these factors are shaped primarily by the prevailing political institutions, which predate resource discovery".[5] In fact, whether derived from natural

---

[2] Ibid., p. 188.

[3] Schmidt-Hebbel, "Fiscal Institutions in Resource-Rich Economies: Lessons from Chile and Norway", *Understanding and Avoiding the Oil Curse in Resource-Rich Arab Economies*, ed. Elbadawi and Selim (2016), p. 228.

[4] Humphreys, Sachs, and Stiglitz, "Introduction: What is the Problem with Natural Resource Wealth?", *Escaping the Resource Curse*, ed. Humphreys, Sachs, and Stiglitz (2007), pp. 8–9.

[5] Elbadawi and Selim, "Overview of Context, Issues and Summary", *Understanding and Avoiding the Oil Curse in Resource-Rich Arab Economies*, ed. Elbadawi and Selim (2016), p. 8.

resources or through other means, wealth can facilitate the implementation of development strategies, many of which have been ambitiously pursued by the states of the Arabian Peninsula.[6]

There are, of course, a number of economic drawbacks to overreliance on a single commodity — in the case of the Persian Gulf states, oil and gas. The oil monarchies, for example, perform far below their potential.[7] Despite good-faith efforts in literally all regional countries to escape oil dependence, its symptoms can be found in three areas: the state and the macro-economy still remain vulnerable to oil price volatility; there is massive overemployment and inefficiency in the public sector; and there is low growth in labor productivity.[8] Other pathologies include boom-bust cycles, dependence on foreign labor, and "a culture of complacency among many of the Gulf nationals, many of whom could easily find 'work' in government".[9]

Nor have most Gulf Cooperation Council (GCC) countries increased non-oil revenues in any meaningful ways.[10] Oil dependence continues to be the norm across the board, with the possible exception of the Emirate of Dubai, although the degree to which the "Dubai model" can survive without the cushion of oil revenues was put to serious test during the 2008 financial meltdown.[11] The Qatari government, for example, has stated that it hopes to finance the country's budget fully from non-hydrocarbon revenues starting from 2020, a goal that seems highly unlikely.[12] Invariably, regional states also remain vulnerable to fluctuations in the oil market. Severe declines in oil prices, especially since 2014, have necessitated deep structural reforms and much-needed diversification, as well as fiscal consolidation and adjustments.[13] The effects of lower oil prices are also being compounded by a number of simultaneous developments, including intra- and inter-state conflicts in Iraq, Syria, Yemen, and elsewhere; serious discord among the member states of the GCC; the further strengthening of the dollar; slowing growth in China and its spillover effects; and recession in Russia.[14]

Given the rapid economic and infrastructural development of the oil monarchies through their deployment of their massive financial resources, and the concurrent improvements they have fostered in the lives of their citizens, it is difficult to point to oil as a catalyst for an economic resource curse in the region. The primary consequences of oil dependence, in fact, have mostly been institutional and political rather than economic. The Persian Gulf's oil states, for example, have benefited from a certain amount of built-in adaptability and enhanced capacity, which has in turn accorded them greater political resilience over time. The petro-states may be suffering from the institutional consequences of resource dependence, but whatever "development" their institutions may have had so far has been sufficient to allow them to ride out, or fend off altogether, crises of the kind that engulfed many of the other Arab states in 2011.

---

[6] As Jeffrey Sachs points out, a successful development strategy needs to have at least three components: public investments suited to national circumstances; a policy framework that supports private sector economic activity; and a political framework to ensure rule of law and macroeconomic stability [Sachs, "How to Handle the Macroeconomics of Wealth", *Escaping the Resource Curse*, ed. Humphreys, Sachs, and Stiglitz (2007), p. 178].

[7] Ibid., p. 174.

[8] Soto and Haouas, "Has the UAE Escaped the Oil Curse?", *Understanding and Avoiding the Oil Curse in Resource-Rich Arab Economies*, ed. Elbadawi and Selim (2016), p. 373.

[9] Foley, *The Arab Gulf States: Beyond Oil and Islam* (2010), p. 144.

[10] Sommer, et al., *Learning to Live with Cheaper Oil: Policy Adjustments in Oil-Exporting Countries of the Middle East and Central Asia* (2016), p. 18.

[11] Davidson, "The Dubai Model: Diversification and Slowdown", *The Political Economy of the Persian Gulf*, ed. Kamrava (2012), pp. 195–220.

[12] Anon., "Qatar: Managing the Limits of Economic Growth", *Gulf States News* 37.956 (2013), p. 11.

[13] Sommer, et al., *Learning to Live with Cheaper Oil*, p. 7.

[14] Ibid., p. 11.

The contrast with other oil states in crisis, for instance such as Libya, is striking. In 2005, the Libyan government attempted to implement subsidy reforms for foodstuffs and energy that had been in place since the 1970s. Food subsidies varied from 39% to 96% of the market price, with flour, rice, sugar, and vegetable oil accounting for the bulk of the state's subsidies. Although the elimination of food and energy subsidies could have saved the Libyan state budget the equivalent of about 2%, it would have cut household expenditure by about 10% and doubled the poverty level.[15] Just before the eruption of the 2011 unrests, the state started to roll back on subsidy reforms, but the measures proved insufficient to stem the tide of the revolution. In the end, despite its best efforts, the state could do little to overcome the cumulative effects of its self-inflicted wounds and to reverse its disintegration.

In the Persian Gulf region, oil may have skewed or impeded changes to political institutions, but it has bestowed on them a certain amount of resilience to withstand or to blunt possibilities for societal pressures for change altogether. It is to an examination of these resource-influenced institutional dynamics that the article turns next.

3   Oil and institutional change

With the exception of Saudi Arabia, formal independence did not come to the Persian Gulf area until 1961 when Kuwait became independent, and even later, in 1971, to the rest of the region. Ruling families, already firmly entrenched in power, now gained unprecedented access to economic resources that enabled them to forge new forms of clientelism and patronage and to deepen existing ones. Meanwhile, the emerging fiscal structures of the petro-states institutionalized a permanent tendency toward rent-seeking, and perpetuated "traditional concepts of authority as the personal patrimony of the ruler".[16] Rentierism became a potent source of legitimacy. In other words, oil shaped subsequent patterns of institutional development, but only in the direction of reinforcing preexisting processes and practices.[17] It also allowed the state to develop new agencies, enhanced its extractive capacities and activities, and shaped the behavior and preferences of policymakers.

When the flow of oil rents and the early stages of the state-building process coincide, rents tend to go directly to the state and are not mediated through domestic private actors. At the same time, economic power and ultimately the political authority of these states resides in their ability to extract rents externally from the global environment and then to distribute the revenues internally.[18] There are, therefore, multilevel and multidimensional causal relationships between oil and institutional development. In the countries of the Arabian Peninsula, as elsewhere, oil revenues made possible the growth and strength of some institutions and the underdevelopment or total absence of others. For example, state leaders were enabled by the flow of oil revenues to excel at taxing foreign oil corporations instead of making unpopular domestic fiscal decisions. They also became good at monitoring, regulating, and promoting the oil industry, both

---

[15] Araar, Choueiri, and Verme, "The Quest for Subsidy Reform in Libya", *The World Bank Policy Group: Research Working Paper* 7225 (2015), pp. 3–4.

[16] Karl, *The Paradox of Plenty: Oil Booms and Petro-States* (1997), pp. 62–3. Contrary to what Karl maintains, there is no evidence to suggest that, at least in the Arabian Peninsula, oil abundance "delayed the development of a modern consciousness of 'the state'" (Karl, *The Paradox of Plenty*, p. 62). In fact, oil appears to have further facilitated such a consciousness by helping political leaders deepen the state's practical and symbolic presence in people's daily lives.

[17] For an accessible history of oil in the Persian Gulf region see: Askari, *Collaborative Colonialism: The Political Economy of Oil in the Persian Gulf* (2013).

[18] Karl, *The Paradox of Plenty*, p. 49.

domestically and internationally, at the expense of the state's ability to develop a coherent, penetrating bureaucracy.[19]

However, successful promotion of the domestic oil industry is far from being the only accomplishment of the Arabian Peninsula's petro-states. In the postindependence era in particular, oil has also brought them a number of other broad advantages, both institutional and otherwise. To begin with, oil can, at the broadest level, offer three huge benefits to oil-producing states: it can boost real living standards by increasing levels of public and private consumption; it can finance higher levels of investment; and, by developing the public sector, it can facilitate state efforts at fostering development.[20] Natural resources can also enhance the state's ability to deliver public goods. In fact, oil-rich states tend to outperform their non-oil neighbors in terms of private consumption and higher levels of well-being, as in decreased child mortality, and increased life expectancy, paved roads, and per capita electricity. The large income generated by oil enables the state to develop capacities that states without oil do not have — e.g., alleviate poverty, become bureaucratically expansive, and hide the extent of their income.[21] Moreover, although rent revenues may foster institutional underdevelopment in certain instances, they are also likely to stimulate state capacity, market capitalism, and industrialization.

We will explore below how, in the context of the Arab petro-states, as in the larger Middle East, rent revenues have undermined rather than enhanced prospects for democracy. But it is hard to deny that rent has also been instrumental in stimulating state capacity, economic growth, and, broadly defined, development. In other words, whether and to what extent rent revenues become a "blessing" or a "curse" depends both on context and on different institutional arenas. There are undoubtedly numerous deleterious economic, political, and institutional consequences of natural resources abundance and the overwhelming reliance on rent revenues accrued from them. But in the strict sense of the term, oil in itself has not been an economic curse.[22] Oil and minerals actually have a positive effect on development. It is not a *resource curse* that causes underdevelopment but an *institutions curse*. States are cursed not by their resources but by their institutions. The economic dependence on natural resources, and the pathologies attributed to it, are caused by weak state capacity, low-quality institutions, and policies pursued by state leaders.[23]

States with weak or inadequate institutions have a hard time imposing and collecting taxes, attracting foreign investments, and otherwise raising revenues. Thus, their dependence on revenues from natural resources is all the more pronounced.[24] Moreover, since states with low-quality institutions cannot create vibrant and diversified economies, natural resources enable them to enjoy the benefits of substantial revenues.[25] In the Arabian Peninsula, albeit under British protection, patterns of rule with tribally rooted systems of patronage and clientelism were already robustly in place before the advent of the oil era. Once rent revenues started flowing into the

---

[19] Ibid., p. 61.

[20] Sachs, "How to Handle the Macroeconomics of Wealth", *Escaping the Resource Curse*, ed. Humphreys, Sachs, and Stiglitz (2007), pp. 174–5.

[21] Ross, *The Oil Curse: How Petroleum Wealth Shapes the Development of Nations* (2012), p. 225.

[22] Ibid., p. 196.

[23] Menaldo, *The Institutions Curse: Natural Resources, Politics, and Development* (2016), p. 11. Menaldo is adamant that "there is no resource curse and oil is not the devil's excrement." Indeed, there is ample evidence that "natural resources wealth improves infrastructure, investments, living standards, helps improve the quality of political and economic institutions, strengthening the state, democracy, and the rule of law" [Menaldo, *The Institutions Curse*, p. 3]. While there is much to be said about the blessings of resource abundance, linking it to democracy and the rule of law may be somewhat of an analytical stretch.

[24] Ibid., p. 18.

[25] Ibid., p. 11.

state, they strengthened rather than altered these patterns, and in the process they also enhanced the autonomy of state leaders.

The key variable is the timing of institutional development in relation to the flow of rent revenues. State leaders can employ rent revenues to strengthen those preexisting institutions on which they rely, and from which they derive their legitimacy. In the Arabian Peninsula, these are the same institutions that perpetuated patronage and clientelism. In Norway and Canada, where democratic institutions were already in place when oil production began, oil enhanced the state's economic performance and helped deepen the state's democratic legitimacy.[26] And, in Chile, where strong financial institutions predated the flow of rents into the economy, what could have been a resource curse was turned into a resource blessing.[27] Some scholars have argued that the political institutions found in the Arab world before the discovery of oil were weak, and that the effects of this weak intuitional setup from the pre-oil era have lingered on.[28] *Weakness* is relative; in the Arabian Peninsula, those institutions from which states emerged — the family, tribal practices and confederations, patronage, and the like — were actually quite strong in relation to the communities over which they began to rule, and subsequently their strength was only enhanced as a result of oil revenues.

Reliance on natural resources wealth does, of course, lead to various negative economic side effects. In the case of oil, there is often unequal expertise between the state and the international oil companies (IOCs), and the latter frequently have leverage over state leaders in terms of information, resources, and access to global markets. There have also been numerous examples of corporations cheating states out of the terms of their contracts, both large and small,[29] and there is also market volatility and unsteadiness of income. Three specific developments can lead to the volatility of income: variations over time in the rates of oil exportation; fluctuations in the value of natural resources; and variability in the timing of payments by the corporations to the states.[30] Resource-dependent states can also be turned into "honey pots" that can be raided by all sorts of domestic and foreign actors,[31] while most states that rely on natural resources income tend to suffer from higher rates of corruption and underinvestment in education.[32] And, as other countries learned the hard way when their natural resources ran out, oil income can easily be used for short-term consumption instead of for longer-term growth and focused investment.[33]

Dependence on single commodity exports shapes classes, regime types, state institutions, decision-making bodies, and decisions related to policymaking.[34] It is also likely to spark off what is known as Dutch Disease: this refers to the condition in which favorable changes in one sector of the economy, such as in the oil sector, cause distortions in other sectors, like

---

[26] Ross, *The Oil Curse*, p. 234.

[27] Schmidt-Hebbel, "Fiscal Institutions in Resource-Rich Economies", p. 225.

[28] Elbadawi and Selim, "Overview of Context, Issues and Summary", p. 3.

[29] Stiglitz, "What Is the Role of the State?", *Escaping the Resource Curse*, ed. Humphreys, Sachs, and Stiglitz (2007), p. 25.

[30] Humphreys, Sachs, and Stiglitz, "Introduction", pp. 5–6.

[31] Karl, "Ensuring Fairness: The Case for a Transparent Social Contract", *Escaping the Resource Curse*, ed. Humphreys, Sachs, and Stiglitz (2007), p. 257.

[32] Humphreys, Sachs, and Stiglitz, "Introduction", p. 10.

[33] Sachs, "How to Handle the Macroeconomics of Wealth", p. 173. Perhaps the most extreme example of a resource-dependent country in which overconsumption led to depletion of the resource on which the economy depended, is that of Nauru, where the once-prosperous Pacific island nation was on the brink of becoming a failed state following the collapse of the country's phosphate industry. See: Connell, "Nauru: The First Failed Pacific State?", *The Round* Table 95.383 (2006), pp. 47–63.

[34] Karl, *The Paradox of Plenty*, p. 7.

agriculture. Income from the export of a natural resource often makes exports from nonnatural commodities more difficult and more expensive, especially in the manufacturing and the agriculture sectors. When revenues accrued from natural resources are used for consumption, as they are in the states that make up the Gulf Cooperation Council, Dutch Disease is a likely outcome.[35] In cases where Dutch Disease has set in, when the market for natural resources slows down, other sectors may find it hard to recover.[36]

In the Arab petro-states, Dutch Disease has manifested itself in declining shares of tradable services, bloated bureaucracies, and a falling-off in agriculture and manufacturing. In the Arab economies, in addition to a general neglect of the agricultural sector, rents from natural resources have led to a diminishing share of services, agriculture, and manufacturing in relation to GDP.[37] In resource-rich Arab countries, it is mostly the domestic production of non-tradable services, such as hotels, restaurants, and real estate, that has grown.[38] This is representative of the rest of the Arab world, where the bulk of the services sector is comprised of travel and transport industries (some 78% in 2008).[39]

4 Political economy of institutional stasis

In addition to its economic consequences, rent-seeking behavior tends to be self-reinforcing and deepens path-dependence, thereby impeding institutional change. This occurs especially during periods of economic boom, when rent revenues are at their height. In oil economies, such boom effects exacerbate petrolization, reinforcing public and private oil-based interests. Political stasis, as will shortly be seen, is also a likely outcome. In rentier states, rulers often maintain support by distributing rent revenues to allies, friends, relatives, and supporters. As one observer has noted, during the first oil boom in the 1970s, oil wealth created an "irrational and symbolic universe" that fostered "societal autism" and helped keep authoritarian systems in place in Algeria, Libya, Iran, and the Arabian Peninsula.[40] Poorer states in particular are more likely to engage in oil exploration and to market it aggressively, as much for political reasons as out of economic necessity.[41]

Reliance on natural resources tends to have negative consequences for institutional and political dynamism. Dependence on oil is particularly harmful to the emergence of accountable, politically representative government. Michael Ross sees oil as a curse because "the revenues it bestows on governments are unusually large, do not come from taxation, fluctuate unpredictably, and can be easily hidden."[42] It is little wonder that petro-states are burdened with a host of difficulties in governance. They often suffer from an information deficit, in the form of lack of feedback mechanisms from citizens to policymakers. In addition, they repeatedly have multiple incentives to develop or employ regulations, and therefore possess the inherent potential for

---

[35] Sachs, "How to Handle the Macroeconomics of Wealth", p. 185.

[36] Dutch Disease can be overcome, or at least lessened, by properly investing oil proceeds as part of a national development strategy. Oil income should be turned into public investment rather than private consumption [Sachs, "How to Handle the Macroeconomics of Wealth", pp. 176–7, 184].

[37] Diop and de Melo, "Dutch Disease in the Services Sector: Evidence from Oil Exporters in the Arab Region", *Understanding and Avoiding the Oil Curse in Resource-Rich Arab Economies*, ed. Elbadawi and Selim (2016), p. 100; Elbadawi and Selim, "Overview of Context, Issues and Summary", p. 6.

[38] Diop and de Melo, "Dutch Disease in the Services Sector", p. 100.

[39] In South Asia, by contrast, it is the ICT and finance industries that predominate in the sector [Ibid., p. 85].

[40] Martinez, *The Violence of Petro-Dollar Regimes: Algeria, Iraq and Libya*, trans. Schoch (2012), p. 41.

[41] Menaldo, *The Institutions Curse*, p. 12.

[42] Ross, *The Oil Curse*, p. 6

overcentralization of power. Finally, they frequently feature few, or at best ineffective, institutional means of connection between citizens and state leaders.[43]

Oil states also tend to spend far more than comparable non-oil states on their military, and are more likely to be involved in conflicts.[44] For instance, in most GCC states, military spending constitutes a higher percentage of GDP than the world average.[45] Since the early 1990s, oil-producing countries have also been about 50 percent more likely to have civil wars.[46]

Additionally, oil states are more likely to be ruled by autocrats, and very seldom do such states report their revenues from oil — for them, political secrecy tends to be the norm rather than the exception. All too frequently, this enables state leaders to divert oil revenues for their own, often corrupt, purposes, lining their own pockets and engaging in cronyism.[47]

Oil abundance can be particularly detrimental to the prospects of democratization. The relationship between oil and democracy is context-specific; if nondemocratic institutions are already in place by the time oil production begins, then changing and democratizing them becomes more difficult. But, as the examples of Canada, the US, and Norway demonstrate, oil on its own does not undermine democratic institutions and practices *if* they are already in place prior to oil's introduction into the economy.

Oil abundance can undermine prospects for democratization through two, interrelated ways. First, oil can impede the emergence of social forces that are likely to press for political participation and accountability. Natural resource extraction can take place without the participation of large segments of the labor force, is often independent of other developments in the economy, and encourages rent-seeking behavior.[48] What has undermined democracy, especially in the Middle East, is not so much the revenues derived from oil, but rather the ways in which oil was discovered and oil pipelines were built, the way oil was transported, and the manner in which its exports and sales were financed.[49] The developments that followed from the production of oil in the Middle East weakened rather than strengthened various forms of political participation. International oil companies collaborated amongst themselves to delay the emergence of an indigenous oil industry in the Middle East, while state leaders saw the control of oil facilitated by overseas interests as an effective means of weakening democratic forces at home.[50]

Once in place, petro-states tend to have oil-based social forces that have a strong vested interest in perpetuating oil-led development. Therefore, any decision by the authorities to build an alternative fiscal base through taxation is very much resisted by the middle classes and the private sector.[51] Timothy Mitchell observes that whereas the age of coal facilitated the birth of democracy by making political mobilization possible, the birth of oil, and the ways in which the global oil industry has grown, has limited possibilities for the emergence of democracy.[52]

---

[43] Karl, "Ensuring Fairness", pp. 264–5.

[44] Humphreys, Sachs, and Stiglitz, "Introduction", p. 13.

[45] Jarzabek, "G.C.C. Military Spending in Era of Low Oil Prices", *MEI Policy Focus 2016–19* (2016), p. 4. In 2016, Saudi Arabia's total defense budget was estimated at $82 billion and was set to rise to $87 billion in 2020, and the UAE's was set at $15.1 billion in 2016 and estimated to reach $17 billion in 2020 [Carvalho, "Gulf States to Maintain Defense Spending Despite Oil Price Slump", *Reuters*, 18 Feb. 2017. See also: Farzanegan's contribution in this issue].

[46] Ross, *The Oil Curse*, p. 145.

[47] Ibid., p. 59.

[48] Humphreys, Sachs, and Stiglitz, "Introduction", p. 4.

[49] Mitchell, *Carbon Democracy: Political Power in the Age of Oil* (2011), p. 253.

[50] Ibid., p. 8.

[51] Karl, *The Paradox of Plenty*, p. 54.

[52] Mitchell, *Carbon Democracy*, p. 254.

Another reason for the inverse relationship between oil abundance and lowered prospects for democratization lies in the enhanced autonomy of state actors. Natural resources are likely to undermine democracy through lessening the government's need for accountability to the citizenry. As unearned income, rent revenues make it more difficult for citizens to mobilize and press the state for accountability. Low taxation as a percentage of total state revenues tends to reduce the incentive for public scrutiny of the government.[53] Rulers and especially autocrats remain in power when citizens believe that governments are delivering benefits relative to their revenues. Since oil revenues are easier to hide, autocrats can boost their popularity by concealing portions of their oil revenues from the public.[54] At the same time, oil revenues also enable states to invest in and enhance their coercive capacities.[55]

Of course, oil wealth does not necessarily stop democratization; it just makes it more difficult. The example of Kuwait, no matter how imperfect its parliamentary system may be, shows that rents do not necessarily make it impossible to constrain the state through representative institutions.[56] However, as Michael Ross observes, "the more oil that an authoritarian government produces, the less likely it will make the journey to democracy."[57]

Even in its most extreme forms, rentierism does not altogether eliminate the possibilities of discontent. Those who are recipients of direct rent transfers often assume they could do better, and the assumption that people can always be bought off politically for the right price has not proved to be universally true.[58] Nevertheless, rent revenues do strengthen state institutions and, more importantly, freeze their development at the stage in which the revenues begin flowing into the coffers of the state. At the very least, such revenues cause the pace at which state institutions change and develop to slow down considerably. In other words, in the case of the Persian Gulf monarchies oil has fostered institutional stasis.

5 Conclusion

On their own, natural resource endowments are a blessing and not a curse. What turns this blessing into a curse is the ways in which it is managed, or, more accurately, mismanaged. For some time, policymakers have known the nature of the remedies for dealing with the resource curse: minimizing the risks of Dutch Disease; instituting policies that enhance growth in short to medium terms; promoting good governance and reducing corruption; preparing for the depletion of oil resource income; and regularly assessing the appropriate policy mix.[59] Long-term savings plans, in the form of investments by sovereign wealth funds, can also help alleviate some of the pitfalls of dependence on and depletion of natural recourses.[60] Of these policy options, the GCC

---

[53] Schmidt-Hebbel, "Fiscal Institutions in Resource-Rich Economies", p. 227.

[54] Ross, *The Oil Curse*, p. 71.

[55] Humphreys, Sachs, and Stiglitz, "Introduction", pp. 12–3. Luis Martinez makes a related argument, maintaining that control over oil revenues enabled former colonies — beset by "a deep sentiment of revenge against the former colonial powers" — to vent their bitterness on their neighboring states, such as Kuwait and Iran in Iraq's case, Morocco in Algeria's case, and Chad in Libya's case [Martinez, *The Violence of Petro-Dollar Regimes*, p. 42]. Warfare, and more broadly the "rally around the flag" effect, make demands for democratization more difficult.

[56] Herb, *Wages of Oil: Parliaments and Economic Development in Kuwait and the UAE* (2014), p. 192.

[57] Ross, *The Oil Curse*, pp. 65–6.

[58] Kamrava, *Inside the Arab State* (2018), pp. 166–70.

[59] Sachs, "How to Handle the Macroeconomics of Wealth", p. 191.

[60] Collier, "Savings and Investment Decisions from Natural Resource Revenues: Implications for Arab Development", *Understanding and Avoiding the Oil Curse in Resource-Rich Arab Economies*, ed. Elbadawi and Selim (2016), p. 286.

states have already made good use of their sovereign wealth funds, and with varying degrees of conviction and success, they have slowly begun to diversify their economies. But on most other scores, including also economic diversification, the political will and the capacity to make difficult decisions seem to be conspicuously absent.

Wealth on its own does not buy state capacity. The capacity that the petro-states have so far enjoyed is rooted in other, complementary phenomena, such as state evolutionary patterns and state-society linkages. But whether these on their own, without the additional element of oil-driven patronage, are sufficient to enable the Gulf states to operate as they have been in the past without substantial changes to their actual institutional setup, is an open question. Within the region, state leaders and opinion-makers frequently talk of the need to be prepared for the inevitable arrival of the post-oil era. With the notable exception of Dubai, little of this talk has so far been translated into reality. How readily the rest of the region can adapt and adopt the Dubai model, and whether by doing so this will impede or expedite the need for other, political changes, is a question that only time can answer.

There are, to state the obvious, differences among Gulf states in terms of patterns of state-building, social cohesion, rent-driven policies, the relationship between ruling families and tribes, and bureaucracies. These differences have been stimulated by implicit understandings between ruling families and other stakeholders, such as the merchants, both before and since the discovery of oil. Clientelism and marital bonds between ruling families and these stakeholders have contributed to the resilience and adaptability of the social contracts between ruling families and stakeholders, and the establishment of enduring ruling bargains.

Further research is needed to explore a number of persisting questions concerning the relationship between resource abundance and institutional evolution in the states of the Persian Gulf. Regional states have invariably responded to economic downturns by promoting nationalization of their labor force, which has decreased the efficiency of state institutions. Some states, however, have taken additional measures to clamp down on any signs of economic or political discontent — most notably, Saudi Arabia, Bahrain, and the UAE — whereas Qatar, Kuwait, and Oman have remained as comparatively benign dictatorships. These variances raise the question of the conditions that shape the responses of these states to oil-induced changes in economic and political conditions. In this particular regard, the entrepreneurial classes deserve special attention. In countries where the state establishes strong clientelistic relationships with the merchant class through offering contracts, the merchant class becomes more dependent on the state, especially during times of downturn. How do these entrepreneurial classes and the state react to different cycles of oil bust and boom? And what does this tell us about institutional evolution in the interim period between the boom and bust cycles?

The articles that appear in this issue address a number of significant questions related to the manifold causes and consequences of the resource curse phenomenon in the Persian Gulf region. Desha Girod and Meir Walters focus on the lingering consequences of colonial legacies for the evolution of what they argue are persistently weak institutions, specifically in relation to Oman and Kuwait. Matthew Gray and Jessie Moritz each explore different aspects of rentierism, with Gray focusing on linkages between rents, neopatrimonialism, and entrepreneurial state capitalism, while Moritz zeroes in on some of the mechanisms that rentier states deploy to co-opt otherwise autonomous social actors. Labor nationalization is one of the main tools used by rentier states to allay social pressures in times of economic downturn, the subject of the article by Gail Buttorff, Nawra Al Lawati, and Bozena Welborne. Finally, Mohammad Reza Farzanegan looks at the impact of oil rents on military spending in the GCC region, exploring what role, if any, is played by corruption. Together, the articles here significantly deepen our understanding of an important but so far understudied phenomenon, namely the manifold consequences of the resource curse in the Persian Gulf.

Bibliography

Anon., "Qatar: Managing the Limits of Economic Growth", *Gulf States News* 37.956 (2013), available online at http://archive.crossborderinformation.com/NetisUtils/Temp/0oflgl45k4vj33u1ayec43ia/fe79e223.pdf?

Araar, Abdelkarim; Nada Choueiri; and Paolo Verme, "The Quest for Subsidy Reform in Libya", *The World Bank Group: Policy Research Working Paper* 7225 (2015), available online at https://openknowledge.worldbank.org/bitstream/handle/10986/21673/WPS7225.pdf?sequence=2&isAllowed=y

Askari, Hossein, *Collaborative Colonialism: The Political Economy of Oil in the Persian Gulf* (New York: Palgrave Macmillan, 2013).

Collier, Paul, "Savings and Investment Decisions from Natural Resource Revenues: Implications for Arab Development", *Understanding and Avoiding the Oil Curse in Resource-Rich Arab Economies*, edited by Ibrahim Elbadawi and Hoda Selim (Cambridge: Cambridge University Press, 2016), pp. 284–300.

Connell, John, "Nauru: The First Failed Pacific State?", *The Round Table* 95.383 (2006), pp. 47–63.

Davidson, Christopher, "The Dubai Model: Diversification and Slowdown", *The Political Economy of the Persian Gulf*, edited by Mehran Kamrava (New York: Columbia University Press, 2012), pp. 195–220.

Diop, Ndiame and Jaime de Melo, "Dutch Disease in the Services Sector: Evidence from Oil Exporters in the Arab Region", *Understanding and Avoiding the Oil Curse in Resource-Rich Arab Economies*, edited by Ibrahim Elbadawi and Hoda Selim (Cambridge: Cambridge University Press, 2016), pp. 82–102.

Elbadawi, Ibrahim and Hoda Selim, "Overview of Context, Issues and Summary", *Understanding and Avoiding the Oil Curse in Resource-Rich Arab Economies*, edited by Ibrahim Elbadawi and Hoda Selim (Cambridge: Cambridge University Press, 2016), pp. 1–15.

Elbadawi, Ibrahim and Raimundo Soto, "Resource Rents, Political Institutions and Economic Growth", *Understanding and Avoiding the Oil Curse in Resource-Rich Arab Economies*, edited by Ibrahim Elbadawi and Hoda Selim (Cambridge: Cambridge University Press, 2016), pp. 187–224.

Foley, Sean, *The Arab Gulf States: Beyond Oil and Islam* (Boulder CO: Lynne Rienner Publishers, 2010).

Herb, Michael, *Wages of Oil: Parliaments and Economic Development in Kuwait and the UAE* (Ithaca NY: Cornell University Press, 2014).

Humphreys, Macartan; Jeffrey D. Sachs; and Joseph E. Stiglitz, "Introduction: What Is the Problem with Natural Resource Wealth?", *Escaping the Resource Curse*, edited by Macartan Humphreys, Jeffrey D. Sachs, and Joseph E. Stiglitz (New York: Columbia University Press, 2007), pp. 1–20.

Jarzabek, Jaroslaw, "G.C.C. Military Spending in Era of Low Oil Prices", *MEI Policy Focus* 2016–19 (2016), available online at www.mei.edu/sites/default/files/publications/PF19_Jarzabek_GCCmilitary_web.pdf.

Kamrava, Mehran, *Inside the Arab State* (New York: Oxford University Press, 2018).

Karl, Terry Lynn, *The Paradox of Plenty: Oil Booms and Petro-States* (Berkeley CA: University of California Press, 1997).

———, "Ensuring Fairness: The Case for a Transparent Social Contract", *Escaping the Resource Curse*, edited by Macartan Humphreys, Jeffrey D. Sachs, and Joseph E. Stiglitz (New York: Columbia University Press, 2007), pp. 256–85.

Martinez, Luis, *The Violence of Petro-Dollar Regimes: Algeria, Iraq and Libya*, translated by Cynthia Schoch (London: Hurst, 2012).

Menaldo, Victor, *The Institutions Curse: Natural Resources, Politics, and Development* (Cambridge: Cambridge University Press, 2016).

Mitchell, Timothy, *Carbon Democracy: Political Power in the Age of Oil* (London: Verso, 2011).

Ross, Michael L., *The Oil Curse: How Petroleum Wealth Shapes the Development of Nations* (Princeton NJ: Princeton University Press, 2012).

Sachs, Jeffrey D., "How to Handle the Macroeconomics of Wealth", *Escaping the Resource Curse*, edited by Macartan Humphreys, Jeffrey D. Sachs, and Joseph E. Stiglitz (New York: Columbia University Press, 2007), pp. 173–93.

Schmidt-Hebbel, Klaus, "Fiscal Institutions in Resource-Rich Economies: Lessons from Chile and Norway", *Understanding and Avoiding the Oil Curse in Resource-Rich Arab Economies*, edited by Ibrahim Elbadawi and Hoda Selim (Cambridge: Cambridge University Press, 2016), pp. 225–83.

Sommer, Martin and et al, *Learning to Live with Cheaper Oil: Policy Adjustments in Oil-Exporting Countries of the Middle East and Central Asia* (Washington DC: The International Monetary Fund, 2016), available online at www.imf.org/external/pubs/ft/dp/2016/mcd1603.pdf.

Stiglitz, Joseph, "What Is the Role of the State?", *Escaping the Resource Curse*, edited by Macartan Humphreys, Jeffrey D. Sachs, and Joseph E. Stiglitz (New York: Columbia University Press, 2007), pp. 23–52.

Soto, Raimundo and Ilham Haouas, "Has the UAE Escaped the Oil Curse?", *Understanding and Avoiding the Oil Curse in Resource-Rich Arab Economies*, edited by Ibrahim Elbadawi and Hoda Selim (Cambridge: Cambridge University Press, 2016), pp. 373–420.

2 Imperial Origins of the Oil Curse

Desha Girod and Meir R. Walters

Abstract: The literature maintains that oil creates a curse on development in countries with weak national institutions at oil discovery, but offers little guidance on the specific institutions that help leaders avoid the curse. We trace rent distribution in Kuwait and Oman, apparent outliers that experienced development despite their weak national institutions at oil discovery. Unlike other examples of the oil curse, Kuwait and Oman contained a strong informal institution that compelled rulers to spend oil revenues on human development: a balance of power between leaders and their domestic rivals. Because informal balances of power are also present in countries with strong formal institutions that avoid the oil curse, this article suggests that the presence or absence of informal balances of power may help account for whether oil is a blessing or a curse.

1 Introduction

The literature on political economy establishes that oil can be a curse or a blessing for socioeconomic development, depending on the strength of national institutions in place at the time of oil discovery.[1] However, the precise institutions that incentivize leaders to spend on socioeconomic development after oil discovery remain unclear. Weak national institutions are generally thought to involve the absence of "formal institutions of government", and, in particular, weak redistributive institutions, power sharing, or checks and balances.[2] But some leaders, particularly in the Arabian Peninsula, who were governing with ineffective formal institutions when they began producing oil, nevertheless decided to invest oil revenues in human and physical capital, leading to remarkable investment in basic services.[3] For example, leaders in Bahrain, Kuwait, Oman, the

---

[1] De Mesquita and Smith, "Leader Survival, Revolutions, and the Nature of Government Finance", *American Journal of Political Science* 54.4 (2010), pp. 936–50; Morrison, "Oil, Nontax Revenue, and the Redistributional Foundations of Regime Stability", *International Organization* 63.1 (2009), pp. 107–38; Robinson, Torvik, and Verdier, "Political Foundations of the Resource Curse", *Journal of Development Economics* 79.2 (2006), pp. 447–68; Luong and Weinthal, *Oil Is Not a Curse: Ownership Structure and Institutions in Soviet Successor States* (2010). An exception is: Stijns, "Natural Resource Abundance and Economic Growth Revisited", *Resources Policy* 30.2 (2005), pp. 107–30.

[2] Amundsen, "Drowning in Oil: Angola's Institutions and the 'Resource Curse'", *Comparative Politics* 46.2 (2014), pp. 171–2. See also: De Oliveira, "Illiberal Peacebuilding in Angola", *The Journal of Modern African Studies* 49.2 (2011), pp. 287–314.

[3] Many scholars have documented an oil curse on development in oil-rich regions, such as the Gulf, but with a different definition of the oil curse. They define it as the lack of either democracy or diversification in

United Arab Emirates, and Saudi Arabia lacked national authority at the time of oil discovery and had to contend with strong tribal societies that controlled territory within their states. Yet these leaders invested in social and economic infrastructure and managed to reduce infant mortality by 87% on average between 1973 and 2012. The steep decline in infant mortality suggests these countries improved health, education, and physical infrastructure.[4]

How did countries with weak formal institutions at oil discovery manage to use oil wealth for broad-based reconstruction? To examine how the Arabian Peninsula avoided the oil curse on human development, the focus here is on two countries in particular: Kuwait and Oman. Like many countries that experience an oil curse, Kuwait and Oman were impoverished, underdeveloped, and lacked effective centralized governments at oil discovery. Their leaders, however, opted for heavy investment of their oil rents on social services and infrastructure. Our analysis emphasizes factors that (1) seem to drive national development in both Kuwait and Oman, and perhaps the Arabian Peninsula more generally; (2) are plausibly present in other regions that experienced the oil blessing, such as Norway and Canada; and (3) contrast with conditions associated with the oil curse in other countries. Of course, there are important differences between Kuwait and Oman. For example, oil was discovered in Kuwait decades earlier than in Oman, and Kuwait was already a unified political entity prior to the discovery of oil. However, Kuwait and Oman faced similar domestic and international constraints that make comparison between them fruitful.

Drawing on archival research as well as the secondary literature concerning both cases, we argue that a crucial factor compelled the leaders of Kuwait and Oman to spend on public programs: a balance of power among rulers and rivals. Prior to the twentieth century, and viewing the region as poor in natural resources and not worth the investment in extractive institutions,[5] the imperial powers left the indigenous balance of power relatively intact. The leaders of Kuwait and Oman thus spent oil revenues on national development to placate potential rivals. In addition, investing in development was favored by the British who were supporting these leaders militarily.

By demonstrating that checks on elite power can exert a significant influence on how resource wealth affects public spending, this study highlights a precise, albeit informal, institution that compels leaders to avoid the resource curse on development. A balance of power is missing in paradigmatic examples of the oil curse, where a dominant group of elites can hoard wealth with impunity. However, a balance of power is present in paradigmatic examples of the oil blessing, where formal power sharing along with checks and balances incentivizes leaders to use rents for human development.

In addition, our analysis suggests that policies of imperial extraction shape governance in oil-rich states today. Countries that lack a history of foreign extraction prior to the discovery of oil retained stronger competing centers of power that shaped broad-based public spending after oil

---

the economy. See, for example: Ross, *The Oil Curse: How Petroleum Wealth Shapes the Development of Nations* (2012). Here, we define it as the lack of investment in basic services and human development. Thus, while all of the cases we discuss may suffer from an oil curse on democracy and economic diversification, they vary in the extent to which they invest in human development.

[4] UNICEF, "Oman: Country Program Document 2012–2015" (2011), p. 2; World Bank, *World Development Indicators* (2015). On infant mortality's correlation with other development metrics, see: Ross, "Is Democracy Good for the Poor?", *American Journal of Political Science* 50.4 (2006), pp. 860–74; Gerring, Thacker, and Alfaro, "Democracy and Human Development", *The Journal of Politics* 74.1 (2012), pp. 1–17.

[5] By extractive institutions, we refer to institutions that "concentrate power in the hands of a small elite and create a high risk of expropriation for the majority of the population" [Acemoglu, Johnson, and Robinson, "Reversal of Fortune: Geography and Institutions in the Making of the Modern World Income Distribution", *The Quarterly Journal of Economics* 117.4 (2002), p. 1235].

was found. Our argument therefore suggests that the oil curse on human development may in fact depend on the history of extraction of other resources by imperial powers.[6]

2 The oil curse

The oil curse is a phenomenon that emerged in the twentieth century, particularly after the 1970s, when leaders took over ownership of their oil production.[7] Oil is associated with a curse on development in two key ways.[8] The first is that oil can reduce the diversification of exports, and the Arabian Peninsula has certainly experienced this version of the curse. This is due largely to the fact that oil is highly profitable for state governments. Because leaders obtain oil rents from the taxes and bonuses they collect from foreign oil companies or from profits that their own national oil companies earn directly, they lack the incentive to invest in the rest of the economy.[9] As a result, the country produces fewer goods locally, and sectors that employ the population at large receive relatively less attention.

In spite of experiencing the first curse on development, countries of the Arabian Peninsula have generally avoided a potential second curse of oil on human development. This second curse emerges when leaders plunder oil wealth at the expense of basic services, national infrastructure, and the general welfare of the population. Leaders are more likely simply to hoard wealth when their power depends on narrow patronage networks and thus lack serious competition from rival groups.[10] Under these circumstances, leaders maximize their wealth by enriching their network at the expense of the rest of the country.

Despite having weak formal institutions of national governance at oil discovery, leaders in countries such as Kuwait and Oman invested newfound oil wealth in broad-based development, rather than simply hoarding it among a small ruling class, as has been the case in West African countries, such as Gabon and Equatorial Guinea, with weak formal institutions. Why did the leaders of Kuwait and Oman choose to spend on national well-being? Put broadly, when countries with weak formal institutions start to produce oil, why do any of them use their windfall revenues to benefit the population at large?

2.1 *Oil for clientelism*

We adhere to the assumption in the political economy literature that leaders' choice to invest in socioeconomic development depends on their strategy to consolidate power. Assuming leaders of states with weak formal institutions who discover oil seek to hang on to power, their primary concern must be the existence of rival groups. Examples of such groups include tribal networks,

---

[6] For a review of the economic history of modern-day development, see: Nunn, "The Importance of History for Economic Development", *Annual Review of Economics* 1 (2009), pp. 65–92. Our argument builds on scholarship on the roles of colonial and indigenous institutions in economic development. See: Acemoglu, Johnson, and Robinson, "The Colonial Origins of Comparative Development: An Empirical Investigation", *The American Economic Review* 91.5 (2001), pp. 1369–401; Arias and Girod, "Indigenous Origins of Colonial Institutions", *Quarterly Journal of Political Science* 9.3 (2014), pp. 371–406.

[7] Andersen and Ross, "The Big Oil Change: A Closer Look at the Haber-Menaldo Analysis", *Comparative Political Studies* 47.7 (2014), pp. 993–1021.

[8] Karl, *The Paradox of Plenty: Oil Booms and Petro-States* (1997); Ross, "The Political Economy of the Resource Curse", *World Politics* 51.2 (1999), pp. 297–322.

[9] Beblawi, "The Rentier State in the Arab World", *Arab Studies Quarterly* 9.4 (1987), pp. 383–98; Ross, "The Political Economy of the Resource Curse", pp. 297–322.

[10] De Mesquita and Smith, "Leader Survival, Revolutions, and the Nature of Government Finance".

religious organizations, factions within a ruling family or ethnic group, military officers, political parties, economic interest groups, or insurgent organizations.[11]

The balance of power among rivals and leaders can, of course, vary considerably within countries with weak formal institutions at oil discovery. Some had clientelistic networks designed to make a single group of elites nationally dominant — economically, politically, and militarily. Such hegemonic concentration of power within one group of elites can emerge either because colonists created it, as in Angola and Equatorial Guinea, or because colonists co-opted power imbalances among indigenous groups in order to profit from the territory, as in Mexico.[12] In particular, extraction of slaves and raw materials led colonists to establish a repressive state apparatus focused on decimating groups outside their local client's immediate network. The hegemony remained intact when oil was discovered; that is, domestic elites who came to power as colonial clients maintained their hold on power and used access to oil strategically so as to maintain and enhance their power. In most former colonies, oil discovery occurred shortly before or after national independence.

At oil discovery, leaders in a dominant political position lacked the incentive to share oil wealth. They could instead use rents to perpetuate the repressive systems of governance they employed under colonial rule. Relying on the existing means of coercion would cost less than mollifying rivals' grievances with broad-based development. In addition, foreign actors seeking to optimize stability for oil production faced incentives to support the ruler regardless of existing repressive tactics because doing so was probably less expensive and risky than helping to buy out the population at large with development projects. Thus, leaders with an apparatus of clientelism and repression already in place were likely to rely on it.

2.2 Oil for development

In contrast, external powers lacked any incentive to micromanage domestic politics in countries that lacked opportunities for substantial extraction during the colonial period.[13] As a result, at oil discovery, no single political faction controlled all of the country's territory. In particular, domestic leaders in states without longstanding extractive networks lacked an established coercive apparatus to maintain national dominance, and had to deal with challenges from rival domestic centers of power. Lacking a dominant position at oil discovery, these leaders faced incentives to invest oil wealth in developing modern national infrastructures and building welfare states as a means to buy off potential domestic rivals.

Specifically, the countries of the Arabian Peninsula — and Kuwait and Oman in particular — suggest that some oil-rich leaders in developing countries faced incentives to invest in redistributive institutions that would appease broad swathes of the population. Below, we first discuss geopolitics in the Persian Gulf, preceding and during the initial years of massive oil production. We then trace the influence of domestic pressure on the leaders of Kuwait and Oman regarding the distribution of oil rents. Both countries lacked the hegemony of a single group of local elites that centuries of intensive colonial intervention had created in other regions, such as West Africa. Unlike quintessential cases of clientelistic petro-states, such as Angola, Gabon, and Equatorial Guinea, competition among rival centers of power in Kuwait and Oman led rulers to use

---

[11] For an example of how indigenous economic interest groups constrained the executive in Botswana, see: Acemoglu, Johnson, and Robinson, "An African Success Story: Botswana", *In Search of Prosperity: Analytic Narratives on Economic Growth*, ed. Rodrik (2003), pp. 80–119.

[12] Arias and Girod, "Indigenous Origins of Colonial Institutions".

[13] Ibid.; Acemoglu, Johnson, and Robinson, "Reversal of Fortune: Geography and Institutions in the Making of the Modern World Income Distribution".

newfound oil revenues on development projects. Moreover, rulers of emerging oil-rich nation-states faced both domestic incentives and pressure from the British to do so as a means of maintaining stability in an age of imperial withdrawal from the region. Because informal balances of power can also be institutionalized into formal systems of power-sharing and checks and balances in democracies such as Norway and Canada that avoid the resource curse, informal institutions may be a necessary condition for avoiding a resource curse on national development. We conclude with implications of our study for research on the resource curse.

3   From poor to rich: the Persian Gulf from 1900 to 1970

For the British, the discovery of oil transformed the Persian Gulf from a region with low strategic importance that they *opted not to* colonize to a region of high strategic importance that they *could not* colonize. The Arabian Peninsula lacked the sorts of natural resources that had attracted colonial rule elsewhere before the oil age.[14] The British were the only Western power actively present in the region for centuries, but instead of using their dominant military capabilities to colonize the Persian Gulf, the British established a Pax Britannica that maintained peace among tribal chiefs. In exchange for the Pax Britannica, the British received control of the chiefs' foreign policy and gained a monopoly on trade on Persian Gulf waters, and thus secured access to the most important British asset in the region: India.[15] But none of these activities was sufficient to incentivize claiming the territories as their own. Instead, the British could achieve their aims in the Persian Gulf without the expense of colonial rule.

The strategic importance of the Gulf increased markedly when the British discovered oil in the region in 1908.[16] Prior to this discovery, Western countries had depended primarily on oil from industrialized countries. But by the early 1960s, one quarter of the world's oil and one half of British oil came from the Persian Gulf, and the percentage seemed likely to increase. As noted by the region's Political Resident from the United Kingdom, William H. Luce,[17] in 1961, "it seems very probable that it [Middle Eastern oil] will have become a vital and irreplaceable factor in the economy and strategy of Western Europe" by 1977.[18]

However, imperial powers were no longer in a position to colonize strategically important territories. During the first half of the twentieth century, the two world wars depleted imperial economies, and international norms favoring decolonization gained strength. Thus, despite oil's importance to the West, colonial expansion was neither logistically feasible, nor in accordance with a growing international norm that favored national self-determination. As a British diplomat noted in 1948:

> The Arab awakening is spreading to these ... kingdoms .... It seems to me that there must be some consistency in our approach .... In our own empire all our efforts are extended towards giving the natives more and more independence. I cannot see how we can hope to move successfully in an opposite direction in the Arab world.[19]

---

[14] Carapico, "Arabia Incognita: An Invitation to Arabian Peninsula Studies", *Counter-Narratives: History, Contemporary Society, and Politics in Saudi Arabia and Yemen*, ed. Al-Rasheed and Vitalis (2004), pp. 11–35.

[15] "Asia & Muscat & Persia & Turkey-Memo. British Interests in the Persian Gulf', 1908, FO 881/9161, p. 3.

[16] "Persian Gulf British Interests in Oil Supply", 1961, POWE 33/2517, pp. 22–23.

[17] Owen, "One Hundred Years of Middle Eastern Oil", *Middle East Brief* 24 (2008), p. 1; "Persian Gulf British Interests in Oil Supply", 1961, POWE 33/2517, p. 27.

[18] "Persian Gulf British Interests in Oil Supply", 1961, POWE 33/2517, p. 18.

[19] "British Advisers in Muscat, Kuwait and Bahrain", 1948, FO 371/68319, p. 7.

As an alternative to obtaining access and stability for oil exploration and production, the British relied on their existing political, economic, and military relationships with local rulers.

While the British were constrained in their ability to colonize the region, they also faced territorial and ideological threats to maintaining their strategic partnerships. As the British Empire collapsed around the world in the 1960s, the Soviet Union promoted communism and pursued alliances with developing countries that would become anti-Western and thus, anti-British. But even more detrimental to the British position in the Middle East was the overthrow by nationalists of the pro-British monarchies. First was Gamal Abdel Nasser who, in 1952, overthrew the pro-British monarchy in Egypt, and inspired Arab nationalism and pan-Arabism. Then, in 1958, Abd al-Karim Qasim overthrew the pro-British monarchy in Iraq, and expressed the view that Kuwait was part of Iraq. Indeed, a 1961 report from the US Department of State indicates that "Arab nationalist ideologies, actively promoted by the Cairo Press and radio, are weakening the fabrics of existing regimes",[20] and that "while the Soviets will undoubtedly support moves to weaken the British position when the opportunities present themselves, the British consider that the immediate threats to their position from the Arab states are of more direct danger".[21]

The British also faced territorial threats from Iraq and Saudi Arabia, both of which seemed to have ambitions to annex their neighbors. Saudi Arabia and Iraq both viewed their neighbors as part of their respective "nations", and also hoped to take over their oil contracts. According to the report from the US Department of State in 1961, "several of the rim states have conflicting territorial ambitions in the Gulf and are waiting to pounce upon the weaker Sheikhdoms in the event of a British disengagement".[22] The US Department of State report highlights Iraq and Saudi Arabia as the "primary threats to current and future British interests in the Gulf".[23]

Instructions in the early 1960s from the Foreign Office to its Political Resident, William Luce, sought to address these concerns. Specifically, London emphasized that it was important "to assist in the steady development of the Persian Gulf states, economically, administratively and politically under the leadership of the present ruling families".[24] By doing so, the British government hoped simultaneously to promote economic development and modernization as well as local ties to Britain. As put by the US Department of State at the time, British policy was seeking "to shift the burden of defense to local states wherever possible".[25]

National leaders in the region, particularly of the small states, were responsive to British advice because they depended on the British for their military power, diplomatic clout, and oil extraction. Indeed, using oil wealth for broad-based development was common throughout the Arabian Peninsula. Leaders across the region used welfare distribution as a strategy for establishing centralized control over new nation-states when they could not rely on indefinite military backing from foreign patrons. As Jill Crystal argues, this was "a patterned, recurring response to oil" throughout the region.[26] In sum, the British advocated national development with oil wealth in the Persian Gulf, and local rulers embraced British advice. The strategy served the interests of both.

In the two following sections, we demonstrate that Kuwait and Oman both experienced low degrees of imperial extraction prior to oil discovery. Consequently, local leaders faced challenges by rival power centers. As a strategy to maintain power, rulers in both countries invested oil rents

---

[20] "Persian Gulf British Interests in Oil Supply", 1961, POWE 33/2517, p. 21.
[21] Ibid., p. 30.
[22] Ibid., p. 21.
[23] Ibid., pp. 29–30.
[24] Ibid., p. 78.
[25] Ibid., p. 31.
[26] Crystal, *Oil and Politics in the Gulf: Rulers and Merchants in Kuwait and Qatar* (1990), p. 2.

to improve the well-being of their populations. While these two oil-rich states did eventually build relatively strong clientelistic and coercive institutions to enrich and protect the power of their royal families, they did so while also building welfare states in which oil wealth was redistributed and invested in extensive national development programs.

4 Kuwait

Following the discovery of oil in the 1930s, Kuwait was transformed from a relatively undeveloped mercantile territory to a wealthy city-state. It now has an extensive network of social services that provides citizens with free education, health care, and social security. Citizens are even guaranteed jobs in the public sector.[27] This outcome depended on a strategic choice made by Kuwaiti leaders to invest significant portions of oil revenues in buying off rivals and promoting national development, rather than relying primarily upon coercion. Hoarding oil revenues and threatening rivals with force would have likely led to instability and violence. Thus, Kuwait's leaders around the time of oil discovery chose to invest rents in national development, rather than contain them among a small group of elites, thereby avoiding a resource curse on development, and hanging on to power.

4.1 *Low imperial extraction enabled domestic political contestation*

During Kuwait's period as a British Protectorate from 1899–1961, the British presence and the level of military incursion into the country were minimal.[28] Before the discovery of oil in the 1930s, Britain's main interest in the country was to ensure that Kuwaiti foreign policy aligned with Britain's strategic interests in the Gulf. On 23 January 1899, the British signed a treaty with Mubarak the Great, the founder of modern Kuwait, as a means of curbing Ottoman influence on the country. The treaty specifically exchanged British military and political support for control over Kuwaiti foreign policy and the power of veto over any foreign concessions, such as railway-building projects, granted by the Kuwaiti government.[29] As a government memorandum in 1908 concerning British interests in the Persian Gulf records, the shaikh signed an agreement that was to be kept absolutely secret, in which he pledged himself "not only to cede no territory, but to receive no foreign Representative without British sanction. In return for this he was promised the good offices of Her Majesty's Government, and a payment was made of 15,000 rupees from the Bushire Treasury".[30] Kuwait's concessions to the British involved maintaining diplomatic efforts with its neighbors in line with Britain's wishes, such as agreeing to draw up its borders with Saudi Arabia according to British terms. Britain thus supplied the tiny Kuwaiti state with military and political support against Ottoman encroachment in exchange for Kuwait's leaders maintaining a pro-British foreign policy, which accorded with Britain's overall strategic aims in the region. However, British involvement in the country's domestic politics was minimal.[31]

---

[27] El-Katiri, Fattouh, and Segal, "Anatomy of an Oil-based Welfare State: Rent Distribution in Kuwait", *The Kuwait Programme on Development, Governance and Globalisation in the Gulf States* 13 (2011).

[28] Crystal, *Kuwait. The Transformation of an Oil State* (1992), p. 12; Askari, *Collaborative Colonialism: The Political Economy of Oil in the Persian Gulf* (2013), p. 12.

[29] Askari, *Collaborative Colonialism*, p. 12.

[30] "Asia & Muscat & Persia & Turkey - Memo. British Interests in the Persian Gulf", 1908, FO 881/9161, p. 17.

[31] Askari, *Collaborative Colonialism*, p. 13. The British did not see Kuwaiti autonomy as a threat since its rulers did not impinge on British maritime interests, and Kuwait was a small and relatively stable state that did not attempt to challenge the regional status quo. See: Crystal, *Kuwait*, p. 12. For further background on

Leading up to the discovery of oil in Kuwait in 1938, there was a balance of power between various political and economic power centers in the country. In other words, the British did not have hegemonic control over Kuwaiti elites as in other contexts where foreign powers reconfigured local politics over centuries. Political and economic power was balanced between the Al-Sabah ruling family and the merchant class that supported the ruling family through financial investment, distribution of British goods, and administering taxes on the pearl trade and shipping.[32] Indeed, the merchant class was so powerful that the Al-Sabah family turned to the Ottomans and then to the British as a strategy to counterbalance the merchants' economic power. However, even with the backing of the British, the royal family struggled to counter the merchant class.[33] For example, there was a tax rebellion in 1909,[34] while an assembly of merchants in 1938 demanded restrictions on the royal family's power and an expansion of investment in social services.[35]

Additionally, significant rifts existed within the royal family itself. For example, Shaikh 'Abdallah, emir of Kuwait since 1950 and during Kuwait's transition to independence from Britain in 1961, eventually exiled his powerful family members, 'Abdallah Mubarak — the only remaining living son of Mubarak the Great — and Shaikh 'Abdallah's brother, Fahad Salim, for siphoning off state funds.[36] Thus, while Shaikh 'Abdallah had initially sought to cultivate ties to the British as a strategy to gain power, ties alone were insufficient to gain dominance over rival domestic groups. Powerful rival indigenous centers of power remained intact but needed to be balanced as Kuwait became an oil-rich independent state.

4.2   Wealth distribution as a strategic choice

Decisions by Shaikh 'Abdallah about how to utilize resource wealth domestically were largely driven by domestic political contestation both within the royal family and between the royal family and the powerful merchant class. The strategy was largely the product of 'Abdallah's attempt to balance power amidst threats from rivals, including from powerful shaikhs within his own family. To centralize political power, 'Abdallah not only used coercive measures such as exiling corrupt family members, but also bought elite support through patronage, expanded development programs, and laid the foundations for representative political institutions.[37] By controlling the production and marketing of oil, the ruling family became the country's largest and most powerful employer while also dominating political and civil society.[38]

The Kuwaiti government bought off elites by purchasing land from members of the royal family and the merchant class at high prices during the 1950s–60s. This land transfer program, which redistributed oil revenues to Kuwait's elite, took up one quarter of the state budget in the early 1960s.[39] In addition to buying off elites from the merchant class and various

---

Kuwait's position in the region and how it fit into British colonial ambitions, see: Askari, *Collaborative Colonialism*; Crystal, *Oil and Politics in the Gulf*; Owen and Pamuk, *A History of Middle Eastern Economies in the Twentieth Century* (1999); Khalaf and Hammoud, "The Emergence of the Oil Welfare State: The Case of Kuwait", *Dialectical Anthropology* 12.3 (1987), pp. 343–57.

[32] Crystal, *Kuwait*, pp. 13–4.

[33] Askari, *Collaborative Colonialism*, p. 13.

[34] Crystal, *Kuwait*, p. 14.

[35] Ibid., pp. 18–9.

[36] Ibid., pp. 25–6.

[37] Ibid., pp. 25–6.

[38] Khalaf and Hammoud, "The Emergence of the Oil Welfare State", p. 351. 'Abdallah also exiled shaikhs with rival policy agendas. Crystal, *Oil and Politics in the Gulf*, p. 12.

[39] Crystal, *Kuwait*, pp. 61–2.

members of the royal family, rulers in Kuwait also bought off the general public by distributing social services and welfare provision more broadly. 'Abdallah initiated an extensive social welfare program for all Kuwaiti nationals, giving them free education, health care, assistance with housing, direct transfers of wealth, and other subsidized goods and services.[40] The ruling family thus put the entire country in the position of its client, dependent upon its patronage, which further cemented perceptions of the family's tremendous power among the broader population.[41]

By investing in national development programs, Kuwait's ruling families could buy public quiescence and consolidate power. As a US Department of State memo noted in 1961:

> The thunder of the reformist-nationalist movement in Kuwait has been largely stolen by the Sabah ruling Family's increasing emphasis on social welfare and modernization. The Government has recently supplemented its already welcome welfare statism with a complete set of modern laws, an up-to-date court system, a cabinet and a constituent assembly .... With earlier-enunciated grievances evidently thus satisfied, no one is very interested in reformist-nationalist agitation. As long as the Sabah ruling family continues on the path of modernization, and as long as Kuwait's wealth continues to be so effectively distributed among Kuwaitis (including nationalists), internal disaffection is highly unlikely.[42]

Although 'Abdallah faced domestic pressure for investing in national development, he also faced external pressure to do so. The success of local attempts to buy political quiescence through national development, along with patronage towards the elite, was in line with British assessments that promoting modernization with oil wealth in countries like Kuwait could promote stability even as the British took an increasingly removed role in supporting local leaders.

By the 1960s, Britain was receiving a majority of its oil from Kuwait,[43] and was expanding British interests in the country from geolocation to include natural resources.[44] This led to attempts to deepen both its political and economic relationships with Kuwait by pressuring Kuwait to use British advisers and make contracts with British companies. Initially, five large British firms won nearly all contracts in Kuwait, and attempted to influence the Kuwaiti government in ways that were more beneficial for the British than they were for Kuwait's economic well-being.[45]

However, even as the British sought to promote policies that would ensure their future access to oil by bolstering their patrons and appeasing domestic rival groups, British officials were conscious that anti-British sentiment could cause any excessive interference in domestic politics to backfire. With major oil production only beginning in the wake of World War II, when the British Empire was already in decline, British officials were aware both of the limits of their influence on Kuwaiti politics and the existence of anti-British sentiment in Kuwait. Given its financial and military limitations amidst the collapse of the British Empire, the British government could not invest heavily in controlling Kuwaiti domestic politics.

---

[40] Crystal, *Oil and Politics in the Gulf*, pp. 10–1; Crystal, *Kuwait*, p. 23; Salih, "Kuwait: Political Consequences of Modernization, 1750–1986", *Middle Eastern Studies* 27.1 (1991), pp. 46–66.

[41] Khalaf and Hammoud, "The Emergence of the Oil Welfare State", p. 331.

[42] "Persian Gulf British Interests in Oil Supply", 1961, POWE 33/2517, p. 41.

[43] Crystal, *Kuwait*, p. 17, 22.

[44] Joyce, "Preserving the Sheikhdom: London, Washington, Iraq and Kuwait, 1958–61", *Middle Eastern Studies* 31.2 (1995), p. 283.

[45] Crystal, *Kuwait*, pp. 24–5; concerning British strategies for using advisers to maximize their influence over Kuwaiti politics without alienating local elites, see: "British Advisers in Muscat, Kuwait and Bahrain", 1948, FO 371/68319, pp. 13–16.

Thus, while the Kuwaiti royal family continued to rely on British support after discovering oil, the British were not in a position to micromanage domestic politics for their clients.[46] For example, as one British official put it in 1954, an "unfortunate but inescapable fact is that in internal matters, and especially in the problem of development, our advice is far from welcome".[47] Similarly, Selwyn Lloyd, the Foreign Secretary, remarked in 1956, that "in Kuwait the Ruler and his family ... resent anything that looks like interference on our part".[48]

British officials also recognized Kuwaiti sensitivity to British interference in external affairs, and the need to preserve Kuwait's national sovereignty.[49] Moreover, given the spread of nationalism, and British worries about the rising influence in the region of both communism and the anti-imperialism of Egypt's Nasser, the British attempted to balance ongoing military support for Kuwait with respect for Kuwaiti national autonomy.[50] For example, according to a memo from the British Foreign Office, it was necessary to consider the limited effectiveness of military arrangements "in keeping the minds of the Kuwaitis oriented our way: if they concluded that their political future lay with Nasser ... and they no longer wanted our military guarantees, there is nothing we could do about it".[51]

In sum, the case of Kuwait illustrates how, following the discovery of oil, domestic and foreign actors could put pressure on leaders to buy off rivals and the general population through a combination of national development and handouts instead of repression and exclusion. In such circumstances, investing in national development became a strategic choice taken by leaders to maximize their chances of survival, while building a nation-state and centralizing power over it. In Kuwait, this probably happened due to the strategic choices of early leaders believing that distributing oil wealth would be the most reliable way to marginalize potential rivals and survive politically. Rivals were present at oil discovery because the British had left competing indigenous centers of power intact.

5 Oman

As in Kuwait, leaders in Oman faced both domestic and foreign pressure to invest in national development after oil discovery. Following the discovery of oil, Omani leaders offered extensive social services in spite of having a longstanding history of weak formal institutions.[52] Thus, Oman serves as another example of a country that used its oil rents for the improvement of human development. Like Kuwait, Oman had strong rival domestic centers of power that were not effaced by colonial intervention before the discovery of oil, which occurred in the 1960s — decades after Kuwait

---

[46] Ashton, "A Microcosm of Decline: British Loss of Nerve and Military Intervention in Jordan and Kuwait, 1958 and 1961", *The Historical Journal* 40.4 (1997), pp. 1069–83; Smith, "The Making of a Neo-Colony? Anglo-Kuwaiti Relations in the Era of Decolonization", *Middle Eastern Studies* 37.1 (2001), pp. 159–72.

[47] Smith, "The Making of a Neo-Colony?", p. 167.

[48] Ibid., pp. 167–8.

[49] As a 1963 letter from the British Ambassador to Kuwait indicates, the British believed that Kuwaiti leaders were aware both of Britain's declining position, and of the need to continue to be seen as independent of excessive imperial influence in order to maintain credibility. Smith, "The Making of a Neo-Colony?", pp. 168–9.

[50] Although Kuwait secured its independence from the United Kingdom in 1961, Iraqi military threats led to a return of British troops shortly afterwards. See: Crystal, *Kuwait*, pp. 91, 126.

[51] "Persian Gulf British Interests in Oil Supply", 1961, POWE 33/2517, p. 3.

[52] According the 2010 Human Development Report, Oman was the country with the fastest increase in Human Development Index globally over the last forty years. See: UNICEF, "Oman: Country Program Document 2012–2015" (2011), p. 2.

and many other countries in the region. And, as in Kuwait, Oman underwent a process of state building shortly after oil revenues began flowing. Sultan Qaboos invested in development as he sought to control an emerging centralized state by buying off potential rivals and gaining acceptance of the Sultanate's newly centralized authority over formerly autonomous regions.

Unlike in Kuwait, the British played a direct and visible role in nation-building in Oman in order to maintain a stable foothold in the geostrategically important region and to secure access to oil before withdrawing in 1971.[53] In their attempts to prevent the spread of anti-British ideologies in the Persian Gulf, so that British interests would be secured without a large foreign military presence in the region, they aimed to stabilize politics through national development that would balance indigenous centers of power and promote stability. According to a British diplomatic memo: "The Sultanate will become increasingly important politically both before and after 1971 as a barrier" to nationalist and communist ideologies.[54] Since the British were planning to depart from Oman, they could neither dictate the decision for local leaders to use oil wealth for human development, nor micromanage how development spending was used. As with Kuwait, local leaders appear to have calculated that they had a strategic interest in using oil rents to buy off rivals and gain popular support, rather than simply relying on coercion to consolidate power. And, as in Kuwait, British interests aligned with those of Sultan Qaboos in Oman.

5.1 *Low imperial extraction enabling domestic political contestation*

British relations with Oman were similar to those with Kuwait prior to the age of oil. Although not a formal British colony or protectorate, Oman was important to the British due to its location along the British Empire's Suez Canal route to India. As also with Kuwait, the British did not try to gain full territorial control over the interior of Oman during this period, essentially maintaining a presence only along the coast.[55] Then, during the early years of oil production, Omani rulers relied heavily on British aid for survival like the Kuwaiti rulers did.[56]

Oman was also similar to Kuwait in its level of domestic political contestation at oil discovery. Power in Oman was divided between an Imamate — with its origins in the eighth century based on Ibadi Islam — which governed the tribes of the interior; and the Sultanate which was based in Muscat on the coast.[57] While the British had aligned with the Sultanate, controlling the interior or assisting the Sultanate in gaining national political hegemony were simply not imperial objectives. Due to this lack of intensive imperial involvement, domestic power struggles persisted throughout Oman prior to the discovery of oil.[58] For example, the Sultanate was challenged during the 1950s by a rebellion from the Imamate (Oman's traditional religious authority), and around the time of the discovery of commercial quantities of oil in Oman in the mid-1960s, a major rebellion broke out in the southern region of Dhofar.[59]

---

[53] "Aid Programme in Persian Gulf", 1967, OD 34/176, pp. 10–11.

[54] "Kuwait-Saudi Arabia Partition of Neutral Zone", 1969, FCO 8/1044, p. 56.

[55] Wilkinson, *The Imamate Tradition of Oman* (1987), p. 274.

[56] Valeri, *Oman: Politics and Society in the Qaboos State* (2009), p. 253.

[57] Wilkinson, *The Imamate Tradition of Oman*, pp. 1–17.

[58] Prior to large-scale oil production, British officials described the Sultan as incapable of controlling his country without direct British support. According to a 1951 British government report on the tribes of Oman, the Sultan had "practically no authority over any of the tribes out of the immediate vicinity of Muscat, Sur, Dhofar, and the British coastal towns. In fact, the tribes of the interior regard themselves ... as completely independent of his authority" ["Oman Tribal Affairs: Reports on Tribes", 1951 FO 1016/37, p. 2].

[59] Worrall, *State Building and Counter Insurgency in Oman: Political, Military and Diplomatic Relations at the End of Empire* (2014), pp. 39–44.

5.2   *Wealth distribution as a strategic choice*

As in Kuwait, leaders in Oman used the distribution of oil wealth as a strategy to consolidate their centralized power over national territory. As also in Kuwait around the time of oil discovery, Oman lacked a single political faction that was nationally dominant economically, politically, and militarily, meaning that aspiring national leaders faced incentives to invest in development as a way to buy off rivals and gain legitimacy among the broader population.

After deposing his father, Sultan Saʿid bin Taimur, Sultan Qaboos used national development as a strategy to buy off potential rivals while attempting to fight an insurgency and build a centralized state. Qaboos modernized Oman by investing oil revenues in major development programs, including hospitals, schools, roads, housing, electricity infrastructure, and in sectors such as agriculture, manufacturing, and fishing. This quickly transformed Oman from an impoverished country into a modern nation resembling other Gulf countries that had discovered oil earlier than Oman.[60]

Britain's own interests in the country coincided with those of Sultan Qaboos. Britain's counterinsurgency strategy in Oman was centered on building support for a newly centralized state through development and social welfare programs.[61] The British viewed investing in development as a crucial strategy, not only for creating a centralized government and promoting long-term stability in Oman, but also for the success of ongoing counterinsurgency efforts.[62] The British reasoned that "in order to leave the Gulf in a stable and peaceful condition and maintain British influence, an entire modern state had to be constructed in the Sultanate almost from scratch".[63] Immediately after the takeover by Qaboos, British government officials judged that popular expectations necessitated the speedy implementation of development projects by the new Sultan,[64] since failure to invest in development could lead not only to insurgencies gaining strength, but also to communist-backed rebels destabilizing the entire region,[65] which would ultimately undermine long-term British military interests and access to oil and trade in the region.[66] Thus, rather than simply assisting Qaboos to defeat the insurgents militarily, the British sought to assist in a broader project of state-building as a means towards long-term

---

[60] O'Reilly, "Omanibalancing: Oman Confronts an Uncertain Future", *The Middle East Journal* 52.1 (1998), p. 73.

[61] Worrall, *State Building and Counter Insurgency in Oman*, pp. 13–4.

[62] According to a 1970 British diplomatic telegraph: "Military measures alone will not pacify Dhofar and must be complemented by civil ones" [1970 "Attitude of Sultan of Muscat Towards Armed Forces", 1970, FCO 8/1414, p. 36].

[63] Worrall, *State Building and Counter Insurgency in Oman*, p. 221.

[64] According to a 1970 British military report: "The coup in July 1970 was received with great popular enthusiasm …. With annual oil revenues of £40 million, the people confidently expected to see rapid improvements to their way of life" ["Relations with Iraq", 1969, FCO 8/1038, p. 37].

[65] According to a British government document dated 11 March 1970: "When I saw the Political Under-Secretary at the Foreign Ministry on 7 March, he expressed great anxiety about reports of the present state of Muscat and Oman …. He said that he felt HMG should do something soon to allay current discontent there: if not, he feared that the Sultan would sooner or later be toppled and that undesirable forces would come into power which would be in an admirable position to upset the existing order further up the Gulf. He referred to the influence of Chinese Communists: and also said that he thought the existence of a future Union of Arab Emirates would be very precarious if Muscat were in hostile hands" ["Political Relations between Muscat and Oman and Iran", 1970, FCO 8/1428, p. 5]. In a 1970 government report on Oman, British officials also assessed the likelihood of Sultan being overthrown by a "'revolutionary' left wing regime" with assistance from China and the Soviet Union ["Rebellion in Dhofar Province", 1970, FCO 8/1415, p. 136].

[66] "Rebellion in Dhofar Province", 1970, FCO 8/1415, p. 90.

national stability.[67] As one British military official put it, "military operations without effective political measures designed to win over the Dhofaris would ultimately be futile".[68]

Indeed, the failure to invest oil money in development as a means to foster national stability was a grievance that British officials had continually raised with Sultan Saʿid bin Taimur. According to a 1970 British military report, attempts to persuade Sultan Saʿid to pursue an active "hearts and minds" campaign had so far "proved abortive. It is, however, vitally important that the uncommitted or non-hostile tribes are won over, and better intelligence obtained about the rebels if the tangible advantages which these additional military forces may gain are not to be lost".[69]

In contrast to his father, Qaboos explicitly embraced the preferred British strategy of centralizing power and defeating challengers, partly through national development and welfare programs.[70] In a press release shared by the British government via diplomatic telegrams on 26 July 1970, Qaboos justified his takeover in terms of his desire to build a modern state with oil revenues: "I have watched with growing dismay and increasing anger the inability of my father to use the new found wealth of this country for the needs of its people. That is why I have taken control".[71] The British tried to persuade Qaboos to undertake a strategy of mixing human development with military force, a strategy that also matched Qaboos's own interests. He publicly promoted himself domestically as a leader committed to development, which ultimately helped him to gain legitimacy and consolidate power. Although it may not be possible to determine the exact degree to which the British or Qaboos initially pushed for investment in development, their incentives fully aligned.

By combining military campaigns with development programs, Qaboos solidified his power and became widely viewed as the legitimate ruler of Oman as a nation. He accomplished this by portraying himself as personally responsible for the development of the state, and thus as personifying the Omani nation as a whole.[72] Overall, like Kuwait, the case of Oman illustrates how resource distribution and investment in social welfare programs can be used by emerging leaders as a strategic choice for nation-building and consolidating power. The British had an interest in stabilizing the Arabian Peninsula. Thus, they had an interest in helping to end civil war in Oman and in centralizing the state. However, as in Kuwait, domestic political pluralism made this difficult to accomplish through force alone, particularly in the wake of the collapse of the British Empire. Using oil wealth for national development was a strategy that served the interests of both the British and the Sultan.

Unlike in Kuwait, where leaders strived to maintain independence from the British in domestic politics, in Oman, the Sultan welcomed visible British participation in wealth distribution. Thus, Oman and Kuwait differed in the degree to which they followed British advice or influence, with the case of Kuwait demonstrating that the British might not have been as essential in driving

---

[67] Worrall, *State Building and Counter Insurgency in Oman*, pp. 14–5. According to a British diplomatic report on technical assistance to Oman: "The Sultanate is so backward and its human resources are so limited that if it is to catch up in the next few years with the development of its neighbors it will need the help of several countries including our own" ["Technical Assistance and Aid for Development to Oman from United Kingdom", 1970, FCO 8/1432, p. 33].

[68] Worrall, *State Building and Counter Insurgency in Oman*, p. 227.

[69] "Rebellion in Dhofar Province", 1970, FCO 8/1415, p. 87.

[70] As a 1970 British government report put it: "The change of leadership has brought about a complete reversal of attitudes towards the economic and social development of the country" ["Technical Assistance and Aid for Development to Oman from United Kingdom", 1970, FCO 8/1432, p. 7].

[71] "Political Developments in Muscat and Oman Following Successful Coup", 1970, FCO 8/1425, p. 100.

[72] Valeri, *Oman*, p. 5.

the emergence of distributive politics as the case of Oman would suggest. Nevertheless, in both countries, distributing oil wealth was in the interest of leaders seeking to build nation-states amidst serious domestic political challenges. Wealth distribution helped leaders in both Kuwait and Oman to consolidate power while preventing violent conflicts from spiraling out of control.

6   Conclusion

The strategic predicament of leaders in Kuwait and Oman led them to invest in national development as a strategic choice to survive amidst rapid domestic and international change. Facing analogous structural pressures, Kuwait and Oman exhibited fundamentally similar outcomes that differed dramatically from quintessential cases of the oil curse in regions such as West Africa that had very different histories of imperial encroachment. Thus, we build on the literature that suggests the resource curse is conditional, affecting only countries that have weak national institutions when they discover oil. By demonstrating that some countries with such weak institutions manage to use resource rents for national development, we highlight domestic political contestation as a variable that compels rulers to promote a strategy of broad-based development in exchange for stability.

The sheer abundance of oil in the Persian Gulf does not explain development outcomes there. In natural resource abundance, Oman is similar to countries such as Equatorial Guinea and Gabon where leaders hoarded oil wealth and left most of their citizens in extreme poverty and without access to basic services. Additionally, the temptation to hoard wealth is also present in the Gulf, as evidenced, for example, by early struggles in the Kuwaiti royal family between factions who wanted to siphon off resources and those who wanted to invest them in development. Thus, development outcomes in the Gulf were not predetermined, but driven, as we argue, by strategic context.

It is important to note that our argument does not focus on the resource curse on democracy. The governments of Kuwait and Oman are autocratic and thus consistent with the resource curse on democracy. Additionally, the mistreatment of foreign workers in many Gulf countries illustrates that even states that spend extravagantly on development are not necessarily concerned with human rights or with all segments of the population.[73] However, the striking development outcomes in countries such as Kuwait and Oman, despite weak national governance at oil discovery, do turn them into cases that contradict the consensus on the resource curse on development, and thus offer an opportunity to investigate new possibilities about the political economy of oil.

Our findings have important implications for theory and for policy. First, the case studies demonstrate that domestic political contestation can serve as an important check on government finances. In other words, when resource-rich leaders need to negotiate with powerful groups in society, the resource curse on development appears to become less likely. Our findings therefore suggest that policies that build up domestic pluralism — by strengthening civil society, for example — can help reduce the oil curse on development.

Second, by demonstrating that imperial and colonial legacies can help to explain why some leaders distributed resource wealth broadly while others used it within a small network of clients only, our study highlights the relevance of indigenous precolonial institutions on modern-day economic development. When extractive imperial institutions destroyed or prevented the emergence of rival domestic centers of power, concentrating resource wealth among a small personalist network was more likely. In countries that did not have a long history of imperial extraction, domestic pluralism and competition for power incentivized rent distribution and the development of social services.

---

[73] Human Rights Watch, "World Report 2018: Kuwait Events of 2017" (2018); Human Rights Watch, "World Report 2018: Oman Events of 2017" (2018).

Finally, and perhaps most strikingly, this article suggests that the oil curse owes its origins to *other* resources; that is, to the possibilities for extraction facing imperial powers during the five hundred years of imperial interventions and colonial rule that preceded the oil discoveries. If domestic political contestation truly shapes whether leaders use oil wealth to benefit the general population, the oil curse may largely depend on the historical legacies of how imperial powers attempted to exploit the pre-oil resource base in different countries.

Bibliography

1 Primary Sources

1.1 *The National Archives (TNA), London, UK*

FO 371/68319: British Advisers in Muscat, Kuwait and Bahrain, 1948
FO 881/9161: Asia & Muscat & Persia & Turkey- Memo. British Interests in the Persian Gulf, 1908
FO 1016/37: Oman Tribal Affairs: Reports on Tribes, 1951
FCO 8/1038: Relations with Iraq, 1969
FCO 8/1044: Kuwait-Saudi Arabia Partition of Neutral Zone, 1969
FCO 8/1414: Attitude of Sultan of Muscat towards Armed Forces, 1970
FCO 8/1415: Rebellion in Dhofar Province, 1970
FCO 8/1425: Political Developments in Muscat and Oman Following Successful Coup, 1970
FCO 8/1428: Political Relations between Muscat and Oman and Iran, 1970
FCO 8/1432: Technical Assistance and Aid for Development to Oman from United Kingdom, 1970
OD 34/176: Aid Programme in Persian Gulf, 1967
POWE 33/2517: Persian Gulf British Interests in Oil Supply, 1961

1.2 *Published sources*

Human Rights Watch, "World Report 2018: Kuwait Events of 2017" (2018), available online at www.hrw. org/world-report/2018/country-chapters/kuwait.
———, "World Report 2018: Oman Events of 2017" (2018), available online at www.hrw.org/world-report/2018/country-chapters/oman.
UNICEF, "Oman: Country Program Document 2012–2015" (2011), p. 2, available online at www.unicef.org/about/execboard/files/Oman_final_approved_2012-2015_20_Oct_2011.pdf.
World Bank, "World Development Indicators" (2015), available online at http://data.worldbank.org/data-catalog/world-development-indicators.

2 Secondary sources

Acemoglu, Daron; Simon Johnson; and James A. Robinson, "The Colonial Origins of Comparative Development: An Empirical Investigation", *The American Economic Review* 91.5 (December 2001), pp. 1369–401, available online at https://economics.mit.edu/files/4123.
———, "Reversal of Fortune: Geography and Institutions in the Making of the Modern World Income Distribution", *The Quarterly Journal of Economics* 117.4 (2002), pp. 1231–94.
———, "An African Success Story: Botswana", *In Search of Prosperity: Analytic Narratives on Economic Growth*, edited by Dani Rodrik (Princeton: Princeton University Press, 2003), pp. 80–119.
Amundsen, Inge, "Drowning in Oil: Angola's Institutions and the 'Resource Curse'", *Comparative Politics* 46.2 (2014), pp. 169–89.
Andersen, Jorgen and Michael Ross, "The Big Oil Change: A Closer Look at the Haber-Menaldo Analysis", *Comparative Political Studies* 47.7 (2014), pp. 993–1021.
Arias, Luz Marina and Desha M. Girod, "Indigenous Origins of Colonial Institutions", *Quarterly Journal of Political Science* 9.3 (2014), pp. 371–406, available online at www.iae.csic.es/investigatorsMaterial/a1217211131378145.pdf.
Ashton, Nigel John, "A Microcosm of Decline: British Loss of Nerve and Military Intervention in Jordan and Kuwait, 1958 and 1961", *The Historical Journal* 40.4 (1997), pp. 1069–83.

Askari, Hossein, *Collaborative Colonialism: The Political Economy of Oil in the Persian Gulf* (New York: Palgrave Macmillan, 2013).

Beblawi, Hazem, "The Rentier State in the Arab World", *Arab Studies Quarterly* 9.4 (1987), pp. 383–98.

Carapico, Sheila, "Arabia Incognita: An Invitation to Arabian Peninsula Studies", *Counter-Narratives: History, Contemporary Society, and Politics in Saudi Arabia and Yemen*, edited by Madawi Al-Rasheed and Robert Vitalis (New York: Palgrave Macmillan, 2004), pp. 11–35.

Crystal, Jill, *Oil and Politics in the Gulf: Rulers and Merchants in Kuwait and Qatar* (New York: Cambridge University Press, 1990).

———, *Kuwait: The Transformation of an Oil State* (Boulder, CO: Westview Press, 1992).

De Mesquita, Bruce Bueno and Alastair Smith, "Leader Survival, Revolutions, and the Nature of Government Finance", *American Journal of Political Science* 54.4 (2010), pp. 936–50.

De Oliveira, Ricardo Soares, "Illiberal Peacebuilding in Angola", *The Journal of Modern African Studies* 49.2 (June 2011), pp. 287–314.

El-Katiri, Laura; Bassam Fattouh; and Paul Segal, "Anatomy of an Oil-Based Welfare State: Rent Distribution in Kuwait", *The Kuwait Programme on Development, Governance and Globalisation in the Gulf States Research Paper* 13, London School of Economics (2011), available online at http://eprints.lse.ac.uk/55663/1/__lse.ac.uk_storage_LIBRARY_Secondary_libfile_shared_repository_Content_Kuwait%20Programme_Fattouh_2011.pdf

Gerring, John; Strom C. Thacker; and Rodrigo Alfaro, "Democracy and Human Development", *The Journal of Politics* 74.1 (2012), pp. 1–17, available online at www.bu.edu/sthacker/files/2012/01/Democracy-and-Human-Development.pdf.

Joyce, Miriam, "Preserving the Sheikhdom: London, Washington, Iraq and Kuwait, 1958–61", *Middle Eastern Studies* 31.2 (1995), p. 281–92.

Karl, Terry Lynn, *The Paradox of Plenty: Oil Booms and Petro-states* (Berkeley: University of California Press, 1997).

Khalaf, Sulayman and Hassan Hammoud, "The Emergence of the Oil Welfare State: The Case of Kuwait", *Dialectical Anthropology* 12.3 (1987), pp. 343–57.

Luong, Pauline Jones and Erika Weinthal, *Oil Is Not a Curse: Ownership Structure and Institutions in Soviet Successor States* (New York: Cambridge University Press, 2010).

Morrison, Kevin M., "Oil, Nontax Revenue, and the Redistributional Foundations of Regime Stability", *International Organization* 63.1 (2009), pp. 107–38, available online at www.pitt.edu/~kmm229/nontaxIO.pdf.

Nunn, Nathan, "The Importance of History for Economic Development", *Annual Review of Economics* 1 (2009), pp. 65–92.

O'Reilly, Marc J., "Omanibalancing: Oman Confronts an Uncertain Future", *The Middle East Journal* 52.1 (1998), pp. 70–84.

Owen, E. Roger, "One Hundred Years of Middle Eastern Oil", *Middle East Brief* 24 (2008), pp. 1–8, available online at www.brandeis.edu/crown/publications/meb/MEB24.pdf.

Owen, Roger and Sevket Pamuk, *A History of Middle Eastern Economies in the Twentieth Century* (Cambridge, MA: Harvard University Press, 1999).

Robinson, James A.; Ragnar Torvik; and Thierry Verdier, "Political Foundations of the Resource Curse", *Journal of Development Economics* 79.2 (2006), pp. 447–68, available online at https://scholar.harvard.edu/files/jrobinson/files/jr_polfoundations.pdf.

Ross, Michael L., "The Political Economy of the Resource Curse", *World Politics* 51.2 (1999), pp. 297–322.

———, "Is Democracy Good for the Poor?", *American Journal of Political Science* 50.4 (2006): pp. 860–74.

———, *The Oil Curse: How Petroleum Wealth Shapes the Development of Nations* (Princeton: Princeton University Press, 2012).

Salih, Kamal Osman, "Kuwait: Political Consequences of Modernization, 1750–1986", *Middle Eastern Studies* 27.1 (1991), pp. 46–66.

Smith, Simon, "The Making of a Neo-Colony? Anglo-Kuwaiti Relations in the Era of Decolonization", *Middle Eastern Studies* 37.1 (2001), pp. 159–72.

Stijns, Jean-Philippe C., "Natural Resource Abundance and Economic Growth Revisited", *Resources Policy* 30.2 (2005), pp. 107–30.

Valeri, Marc, *Oman: Politics and Society in the Qaboos State* (New York: Columbia University Press, 2009).

Wilkinson, John Craven, *The Imamate Tradition of Oman* (Cambridge: Cambridge University Press, 1987).

Worrall, James J., *State Building and Counter Insurgency in Oman: Political, Military and Diplomatic Relations at the End of Empire* (London: I. B. Tauris, 2014).

3   Rentierism's Siblings

On the Linkages between Rents, Neopatrimonialism, and Entrepreneurial State Capitalism in the Persian Gulf Monarchies

Matthew Gray

Abstract: This paper examines rentier state theory (RST), and specifically "rentierism" as a more refined and nuanced variant of RST, arguing that while rentierism provides considerable utility in explaining the state-society relationships of the contemporary Arab states of the Persian Gulf, it is insufficient as a stand-alone explanation, and needs to be considered as a political dynamic of the state-society relationship, rather than as a structural explanation for the state itself, as early RST more ambitiously sought to do. Rentierism therefore needs to be utilized in combination with two other explanatory frameworks, neopatrimonialism and state capitalism. In effect, these are rentierism's theoretical "siblings": they sharpen a rentier analysis by providing greater nuance about how elite networks, business-government relations, and personalized politics operate and interact in the allocative settings of the Gulf, as well as illustrating both the scope and the limits of rentierism as an explanatory framework.

1   Introduction

Rentier state theory (RST) argues that states with large external incomes from sources such as oil and gas royalties do not face the same imperative to offer society democratic and redistributive bargains, compared with states that must impose taxation on their citizens. It is perhaps the most substantial theoretical contribution that Middle Eastern studies has made to the discipline of political science. In various forms, it remains virtually a default assumption about the political economies of the Arab monarchies of the Persian Gulf, notwithstanding some important works that seek to substantially refine the concept,[1] challenge its assumptions of "negative prognostications" for such states,[2] or look beyond it for evidence of societal agency.[3]

What is not usually appreciated by these works, however, is the evolution of the theory since its early manifestations, initially by Mahdavy and subsequently by Beblawi and Luciani.[4] Indeed,

---

[1] Gengler, *Group Conflict and Political Mobilization in Bahrain and the Arab Gulf* (2015).

[2] Springborg, "GCC Countries as 'Rentier States' Revisited", *Middle East Journal* 67.2 (2013), p. 301.

[3] One example where this is done well is: Foley, *The Arab Gulf States: Beyond Oil and Islam* (2010).

[4] The first piece to use the concept as it is used now was Mahdavy, "The Patterns and Problems of Economic Development in Rentier States: The Case of Iran", *Studies in Economic History of the Middle East*, ed. Cook (1970), pp. 428–67. The concept only gained traction among scholars later, with the publication of Beblawi and Luciani (eds), *The Rentier State: Nation, State and the Integration of the Arab World* (1987). Thus, Beblawi and Luciani are most typically credited with developing the concept.

RST can be seen to have moved through at least two major phases of development, in the process of which it has been restructured and honed from its initially overly optimistic goal of being an explanation for state *structures*, into the more realistic aim of *rentierism* explaining a political *dynamic* and strategy used by states in oil-rich political economies.[5] When Gengler writes that "political life in the Arab Gulf states can no longer be summarized neatly and axiomatically as a pragmatic bargain of economic happiness for political quietude",[6] he is correct: not because RST is conceptually invalid, but because Gengler is critiquing the older, more ambitious and simplistic manifestation of the theory, rather than considering allocative politics as a (rentier) *dynamic*. Understood as the latter, the concept of rentierism remains a valid and, in fact, convincing concept, still integral to understanding the political impacts of natural-resource wealth in the Persian Gulf states and elsewhere.

However, a reformulation of the scope of rentier theory means that it cannot act in isolation as a comprehensive explanatory framework: it must be situated in and interrelated to other dynamics, including elite politics, economic structures, and — as much as it was denied by the early literature — even societal agency and activism. This paper explores the main linkages between rentierism and other theoretical explanations to illustrate the continued utility of the rentier concept. It also acknowledges both its limitations and the scope it offers for a variety of non-state actors and forces to have agency and voice, even where rentierism is the prime mechanism through which the state pursues popular acquiescence and regime durability. There is, in other words, no need for mutual exclusivity between rentierism and certain other explanations. Specifically, it is argued that there is a tripartite dynamic between rentierism, neopatrimonialism, and the Gulf monarchies' unique form of "new" state capitalism, in which these three dynamics support and reinforce each other. While each of the three can be examined independently, to a certain extent, they provide the greatest explanatory efficacy for the contemporary Arab states of the Persian Gulf when considered in conglomeration.

This approach goes a considerable way towards addressing many of the questions and criticisms that have been directed at rentier explanations. How was it that the shah of Iran was overthrown, despite the enormous external support, repressive capacity, and, above all, the tremendous oil wealth, all at his disposal? How is it that Kuwait is among the most rentier-directed political economies in the region, but contrary to rentier expectations, also has the most activist parliament? Why do even the richest states with largest endowments of oil and gas, despite early rentier predictions, so actively pursue development strategies, economic diversification, and new employment opportunities for their nationals — even when their hydrocarbon income is set to last many more decades, presumably allaying any need to rush to reform? The answers to these questions demonstrate not that rentierism does not exist as a political dynamic, but that earlier versions of rentier state theory failed to build a sufficiently sophisticated and nuanced explanatory framework. By linking rentierism to neopatrimonialism and state capitalism, the aim here is to go some of the way towards refining and improving rentier theory, while highlighting the importance of neopatrimonialism and (new) state capitalism to an understanding of Gulf politics, especially since they have been underemphasized in the literature to date.

The tripartite framework outlined here argues for a partial overlap of the three dynamics and relationships in each direction. While rentierism is always present in the contemporary Gulf monarchies, new state capitalism or neopatrimonialism may, at times, be more powerful or salient than the other(s), or potentially even more important for regime maintenance and stability than the

---

[5] Gray, "A Theory of 'Late Rentierism' in the Arab States of the Gulf", *Center for International and Regional Studies Occasional Paper* 7 (2011).

[6] Gengler, *Group Conflict and Political Mobilization*, p. 153.

rentier dynamic. More typically, however, factors such as the historical basis of the state-society relationship, the nature of the business-government relationship, or the strength of tribal, religious, provincial, or other elites, will create a long-term arrangement wherein rentierism will always be a core, if broad, co-optive mechanism, but with the importance of state capitalism and neopatrimonialism varying from state to state. To preempt later discussion, this is why Kuwait can be among the most rentierist of the region's political economies but also have the most powerful and activist parliament of the Gulf monarchies.[7] The social origins of the ruling Al Sabah family, who come from the same social structures as the main commercial and tribal elites, and the enduring importance of tribal and sectarian affiliation, mean that the royals must consult with these elites and maintain their confidence. Thus, neopatrimonialism and the relative power of the merchants are stronger in Kuwait than in some other Gulf states. Somewhat in contrast, the Qatari royal family can operate in a smaller clique, maintaining power by a stronger emphasis on blunt rentier allocations. State capitalism and neopatrimonialism are of course important in Qatar too, but to a markedly lesser degree than, for instance, in Kuwait.

The idea of "neopatrimonialism" has been discussed in the context of rentierism in the past, if often briefly or implicitly.[8] It is defined here as an interpersonal mechanism wherein a leader creates a web of elites around himself — the ruler is usually a male — and encourages those elites to form their own patron-client networks through which resources and opportunities are fed out by the ruler, and loyalty and information are fed inward. Provided that some rivalry or mistrust is maintained between key elites, this is a very effective means by which a ruler or a small coterie of royals or elites can dominate an extensive political system. Alone it is usually insufficient to maintain public loyalty or ensure regime legitimacy, but when married with rentierism it becomes a much more formidable political strategy. In contrast with neopatrimonialism, "new state capitalism" is a more recent concept, but is based on the older concept of "state capitalism". Ian Bremmer and others began to discuss it at the beginning of the twenty-first century.[9] It is meant here as a political economy in which the state is a disproportionate owner of the means of production in an economy, and is highly regulatory and paternalistic towards the private sector, while nonetheless permitting market forces to set prices and usually encouraging state-owned firms to be efficient and profitable. The state has long-term goals and is even "entrepreneurial" in its ambitions.[10] The state is not strongly driven by distributive justice or ideological concerns, as in earlier state capitalism models, but seeks to expand the resources and economic reach of the state for the purposes of enhancing its legitimacy and political control.

Rentierism, it is argued, has "siblings" in two metaphorical senses of the term. The first, as already implied, is that as a theory it has other theoretical explanations — in this case, neopatrimonialism and new state capitalism — to which it is inextricably tied. They are siblings by virtue of their interconnectedness and their reliance on each other: the funds from rents help establish and sustain state capitalist enterprises, while the elite bargains and patron-client networks of

---

[7] Among the six Gulf Cooperation Council (GCC) states, Kuwait is *the most* rentierist if this is measured by oil as a share of GDP or as a share of exports, and ranks lowest for export diversity. For rents as a share of government revenue, it falls around the middle of the six. See: International Monetary Fund (IMF), *Economic Diversification in Oil-Exporting Arab Countries* (2016), pp. 8, 10, 12–13.

[8] Some cases where it is discussed in some detail include, among others, the mostly historical discussion in: Hertog, *Princes, Brokers, and Bureaucrats: Oil and the States in Saudi Arabia* (2010); a brief discussion of the Saudi private sector's role in: Niblock and Malik, *The Political Economy of Saudi Arabia* (2007); and cases from Qaboos's Oman in: Valeri, *Oman: Politics and Society in the Qaboos State* (2009).

[9] Bremmer was not the first to identify the dynamic, but was among the first to examine it as a global twenty-first century trend and to term it a new variety of state capitalism. For his main work on the subject see: Bremmer, *The End of the Free Market: Who Wins the War between States and Corporations?* (2010).

[10] Gray, "A Theory of 'Late Rentierism'", pp. 34–5.

neopatrimonialism are both distributive pathways for rents and a clientelist interpersonal structure through which entrepreneurial state capitalism is ordered and nourished. While the mechanisms of these dynamics may be usefully examined without reference to multiple theories, a more holistic approach offers greater scope for understanding how and why the politics of the Gulf monarchies are so often personalized and opaque — and why rents remain central to Gulf politics.

The term "siblings" is also a conceptual reference to the political actors and forces beyond the state; ones that engage with rentierism but that also appear in neopatrimonial webs and state capitalist settings. As is argued here, the head of a state-owned firm is at once a state capitalist and, typically, both a patron and a client in the neopatrimonial political structure, as well as a direct and indirect beneficiary of the allocation of rents. As another example, a similar argument can be made for a merchant who is politically connected and whose commercial interests link to state expenditure or consumption. While there may sometimes be merit in examining these actors in and of themselves, and sometimes from a single explanatory perspective, it is folly to examine them through too narrow an approach when seeking to explain political economies more broadly.

This paper starts with the assumption that the primary and ultimate outcome sought by all ruling regimes is, first and foremost, political survival. This should not be a contentious point of departure. Moreover, where possible, regimes also seek to thrive politically, not only to sustain their rule but also to build and enhance their legitimacy and popularity. In so doing, it is proposed that the Gulf monarchies have six specific goals in mind when pursuing these two main outcomes:

1. To ensure popular acquiescence towards the ruling elite and the political order;
2. To avoid civil society gaining autonomous power to a point where it could mount effective opposition to the regime;
3. To manage elite relationships, co-opt key elites, and avoid elites being able to mount opposition to the regime or to contest it effectively;
4. Partly in service to 1 and 2 above, and partly in and of itself, to influence and be able to influence communications pathways between — and dialogues among — state, civil society, and society, including both state-society discourse mediated by civil society as well as direct communication between state and society;
5. To manage the operation of the economy, especially the distribution of oil and gas wealth but also other "strategic" sectors, while also diversifying the economy, reforming it if and when necessary, and managing the benefits of economic development; and
6. To expand the ruling elites' popular legitimacy and support, but cautiously and only if and when it is judged politically safe to do so.

The three approaches that are the focus here have extensive and deep relationships with *all* of these goals. Rentierism is central to nearly all of them; only with the fourth goal is there not such a strong link to rentierism, but even then, rents pay for both the repressive and co-optive tools used by regimes towards that aim. In all other cases, rents pay for direct co-optation and repression, or indirectly support such aims. Likewise, new or entrepreneurial state capitalism is integral to most of these goals as well; only indirectly to the first one, perhaps, but otherwise central to how the state manages elite relations, civil society, and many aspects of its engagement with society. Finally, neopatrimonialism is crucial to these goals as well. While it is not directly central to the state's engagement with society as a whole, neopatrimonialism is still a mechanism through which the state links to and communicates with society. In so doing, rulers and other political elites build extensive networks with elites outside the core political system: with key families, tribes, businesspeople, religious elites, cultural producers, and others. Indeed, they must do so in order to survive, for even though rentier wealth bestows enormous capabilities and advantages on extant leaderships, rents *alone* are insufficient to ensure an incumbent regime's survival. It is a combination of rentierism, entrepreneurial state capitalism, and neopatrimonialism that creates the formidable ruling capacity that characterizes the Gulf monarchies.

2   From rentier state theory to (late) rentierism

Rentier State Theory and its close variants — including what here is called "rentierism", to distinguish it as a political dynamic rather than a structural explanation for a political economy — is now a common explanatory framework used by analysts of oil-rich economies, including those writing on the Gulf. Yet it is not without its critics. Taken as a comprehensive *structural* explanation for oil-centered political economies, it is indeed a problematic thesis because of its oversimplification of complicated political and social dynamics. The early rentier literature suffered from excessive ambition and oversimplification: Luciani's and Beblawi's early work, for example, argued too simplistically for a bipolarity of either: "exoteric" (or "allocative") states, where rents are received by the state from abroad and allocated to society, or "esoteric" or ("production") states that tax domestic productive activity and redistribute that money.[11] The conclusion from this dichotomous division is that exoteric, or rentier, states have no reason to raise taxes, and therefore no reason to make democratic bargains with society.[12] Such reasoning is problematic, as the example of Kuwait, already mentioned, illustrates.[13] Other problems with early RST were claims that the state would be autonomous from society, and that the state would see no need for an economic or development policy apart, perhaps, from an expenditure policy.[14] Such claims by RST advocates were demonstrably invalid: more than a decade before RST emerged, Saudi Arabia's first development plan covered the period 1970–75, showing that even in the early rentier period, states still concerned themselves with issues of development and economic management. Moreover, a glance at any Gulf state will yield examples of social forces exercising agency.[15]

There is more to state-society interaction than a one-directional economic-allocative link. It is convincing, however, as a less ambitious explanation for political *dynamics*, and for the ways in which oil shapes and skews political *processes*, where allocation has been, and remains, a key mechanism of state control over elites and broader society. Perhaps because the initial work by Luciani and Beblawi was flawed but seemed to contain considerable promise, various attempts to refine and sophisticate their work followed. Broadly, rentier theory itself was subjected to greater scrutiny. Some of its less plausible aspects were abandoned, and most of its key arguments refined and fleshed out; comparative work by Crystal, Chaudhry, and Moore, and Hertog's study of Saudi Arabia, are examples.[16] RST was also enriched by being combined with other theories

---

[11] Both Luciani and Beblawi made such arguments; a good summary of the idea is provided in: Luciani, "Allocation vs. Production States: A Theoretical Framework", *The Arab State*, ed. Luciani (1990), pp. 70–4.

[12] Luciani, "Allocation vs. Production States", pp. 75–6.

[13] On this question of rentierism and Kuwait, see: Herb, *The Wages of Oil: Parliaments and Economic Development in Kuwait and the UAE* (2014), pp. 9–15, 60–3, 158–67, 206–10; and Crystal, *Oil and Politics in the Gulf: Rulers and Merchants in Kuwait and Qatar* (1995), pp. 109–11, 178–83, 201–4.

[14] As was argued in, for example, Luciani, "Allocation vs. Production States", p. 76.

[15] On the plan and others that followed from 1970 to 1985, see: Niblock and Malik, *The Political Economy of Saudi Arabia*, pp. 52–72.

[16] Crystal, *Oil and Politics in the Gulf*; Chaudhry, *The Price of Wealth: Economies and Institutions in the Middle East* (1997); Moore, *Doing Business in the Middle East: Politics and Economic Crisis in Jordan and Kuwait* (2004); and Hertog, *Princes, Brokers, and Bureaucrats*.

and approaches, including international relations theories and sociological arguments,[17] and by being placed into historical contexts by Herb, Al-Rasheed, and others.[18] It was also combined with other arguments from political economy, such as business-government relations, comparative sectoral studies, and economic liberalization debates. Such work is not dissimilar in approach to what is being attempted here, with a modified rentier argument being combined with an explanation from political science (neopatrimonialism) and another from political economy (new state capitalism).

The argument here is that the Gulf should be seen as having entered a period of "late rentierism".[19] This concept defends and refines the fundamental tenets of rentierism, but argues that the political economies of the Gulf became more complex as a result of the oil booms and busts from the 1970s. States became more politically sophisticated, and the state-society relationship was changed by technological transformations, population growth, and the need to prepare for the end of the hydrocarbon era. Late rentierism agrees that rents do indeed make democratization much less likely than would otherwise be the case, but argues that states must still be somewhat *responsive* to society. States pursue economic and development goals, in contrast to what early RST argued, but as a way to tinker with rentier settings, not towards genuine economic — much less political — transformation. Rulers in rentier settings will open their political economies to globalization, but seek to do so on their own terms. Late rentierism also highlights why these states pursue active foreign policy goals and why they are long-term in their thinking. Others have touched on some of these arguments,[20] and further research has examined other elements such as globalization or new state capitalism.[21]

To argue the converse, that rentierism should be abandoned as an explanation, is unconvincing. While the theory may be fluid and changing, this reflects the rapid changes taking place in the Gulf, and the ambitious nature of the theory, not the fundamental weakness of the theory. The sheer size of the rents flowing into the region, and the unique distributive mechanisms through which they pass, means that at least *some* discussion of them is essential for any comprehensive study of the Gulf. The weakness of political institutions and formal opposition forces in the region also supports some of the basic rentier ideas. More recently, the centrality of rent income to state development strategies, infrastructure, and rivalries between Gulf states all further suggest that rentier approaches cannot be dismissed. This approach does not necessarily negate others, such as Herb's categorization of "extreme" rentiers (Kuwait, Qatar, the UAE), "middling" rentiers (Saudi Arabia, Bahrain, Oman, and others), and "poor" rentiers (Iran, and a range of other, relatively poor states).[22]

That said, late rentierism was specifically formulated with the "extreme", and to a lesser extent "middling", rentiers firmly in mind. Herb's categorization only modifies the approach here insofar

---

[17] As an example of a work that combines international relations and RST, see: Gause, *Oil Monarchies: Domestic and Security Challenges in the Arab Gulf States* (1994). There is a range of works that accept some basic validity with RST, but which argue that other approaches are also essential. As examples, see: Foley, *The Arab Gulf States*; and some of the pieces in: Al-Rasheed and Vitalis (eds), *Counter-Narratives: History, Contemporary Society, and Politics in Saudi Arabia and Yemen* (2004).

[18] Herb, *All in the Family: Absolutism, Revolution, and Democracy in the Middle Eastern*; Al-Rasheed, *A History of Saudi Arabia* (2010); Jones, *Desert Kingdom: How Oil and Water Forged Modern Saudi Arabia* (2010).

[19] What follows is drawn from: Gray, "A Theory of 'Late Rentierism'".

[20] On some of the economic points see, Ross, *The Oil Curse: How Petroleum Wealth Shapes the Development of Nations* (2012); and, Hertog, *Princes, Brokers, and Bureaucrats*, which refined the concepts of "state" and "society".

[21] See, for example: Fox, Mourtada-Sabbah, and al-Mutawa (eds), *Globalization and the Gulf* (2006); Marcel and Mitchell, *Oil Titans: National Oil Companies in the Middle East* (2006).

[22] Herb, *The Wages of Oil*, pp. 10–15.

as "middling" rentiers will, by definition, have lower rents per capita than "extreme" ones. This *may* tilt the political scales away from rentierism and towards neopatrimonialism or new state capitalism, but not necessarily. Arguably Qatar and the UAE, especially the emirates of Dubai and Abu Dhabi, are the most entrepreneurial of the Gulf states in their use of state capitalism, and are also less reliant on neopatrimonial dynamics than most others. Rentier theory continues to demand greater refinement and adjustment, including combination with other explanations.

The only way in which the overarching allocative structure of these political economies is likely to change will be if there is a marked and urgent increase in state capacity, or in their legitimacy from non-economic sources. Such a change in capacity might be thrust upon the region if hydrocarbons, or international demand for them, run out very quickly. This could present regimes with the opportunity to restructure state-society relations, while deflecting responsibility for any new hardships away from the state. States may also, or alternatively, try to build their legitimacy through non-economic means, such as by strengthening the role of national mythologies; negotiating deep social change on condition that there is no political transformation; or stoking social tensions. The latter of these has already occurred in some states, such as in Bahrain, where the regime has overstated the threats of sectarianism and arguably used sects as a political strategy.[23] However, such maneuvers are risky. They draw parts of society closer to the ruling elite, and may boost regime claims to power, but carry the risk of greater societal conflict which, if realized, can instead undermine the legitimacy of the incumbent elite.

3   Elite politics, neopatrimonialism, and rentierism

The first explanation with which late rentierism can be combined is neopatrimonialism. Doing so helps describe and explain the more nuanced and subtle mechanisms of power than state allocation and passive societal acceptance of a rentier "bargain". Neopatrimonialism extends the Weberian idea of "patrimonialism" to modern bureaucratic and state institutions and structures.[24] The original patrimonialism concept was that a central ruler held overwhelming dominance, building a network of supporters through which favors would flow in exchange for loyalty and support. For Weber, patrimonialism was akin to family politics, with power centered on and flowing from a male patriarch, but conducted on a larger (tribal, village, or district) stage. Little or no authority was needed to be placed in external institutions or norms, he argued, even if small, highly informal bureaucracies could exist. Later analyses developed the concept, often in studies of African politics, where it was seen as a useful way of explaining the problems of informal politics, corruption, and economic underperformance.[25] Neopatrimonialism sought to explain the endurance of such traditional politics in the modern bureaucratic politics of states.

Eisenstadt reinvigorated scholarly debate about patrimonialism in general by introducing the idea of neopatrimonialism, or modern patrimonialism;[26] other scholars then developed the idea further. A precise definition of neopatrimonialism remains somewhat elusive, although Clapham's definition is widely accepted as:

> a form of organization in which relationships of a broadly patrimonial type pervade a political and administrative system which is formally constructed on rational-legal lines. Officials hold positions

---

[23] Wehrey, *Sectarian Politics in the Gulf: From the Iraq War to the Arab Uprisings* (2014), pp. 58–72.

[24] Outlined in his posthumously published work: Weber, *Economy and Society* (1922; repr. 2017).

[25] This has been widely noted, but see, for example: Mkandawire, "Neopatrimonialism and the Political Economy of Economic Performance in Africa: Critical Reflections", *World Politics* 67.3 (2015), pp. 563–612.

[26] Eisenstadt, *Traditional Patrimonialism and Modern Neopatrimonialism* (1973).

in bureaucratic organizations with powers that are formally defined, but exercise those powers, so far as they can, as a form not of public service but of private property.[27]

This broadly explains both the informality and personalization of neopatrimonial systems and the blurring of public and private sectors. More specifically, in neopatrimonial systems as in traditional patrimonial ones, elites are recruited not (just) for competency but also for their loyalty to the ruler. Thus, their authority stems less from their positions than their distance from the ruler.[28] Elites also compete with each other for access to the ruler, resources, and new opportunities. Indeed, the most effective neopatrimonial rulers deliberately stoke such rivalries to ensure that their clients do not combine against the ruler, to keep them dependent on largesse and favors from above, and so that they rely on the ruler to act as a referee between them.[29]

Elites then develop their own patron-client networks, extending their own reach down into the large political institutions and social groups in which, at various levels, they are representing the ruler and the regime, and which they are managing and reporting upon. These relationships typically also incorporate intermediaries or brokers, who may act as a link between the more senior and busier patron and the lower-level client.[30] Through these multi-layered patron-client networks, a ruler can, albeit indirectly, have knowledge of, and power over, large political units, even entire polities. When neopatrimonialism is managed well, a ruler's autocratic power "penetrates society in a particularly deep manner",[31] creating considerable durability through what Heydemann called "networks of privilege".[32] The delicacy of such networks increases with their complexity, to be sure, but where the relationships are managed effectively and resources are used adroitly, the symbiosis between patron and client, at every level, fosters in them enormous stability. However, such a system also reduces the scope for systemic reform, even when this is urgent, since reform almost invariably undermines the patron-client structures on which the ruler's power depends.[33]

Neopatrimonialism is integral to contemporary Gulf politics, where power is tightly centered on the ruler and an inner elite, the latter usually consisting in substantial part of royals and members of well-connected major families. Patron-client networks, including elite pluralism, competition, and balance, connect that inner elite to institutions and social groups, including the military and bureaucracy;[34] business, both large merchant families and smaller businesspeople; and state-owned firms.[35] Meanwhile, society is kept "politically inchoate",[36] although Gulf

---

[27] Clapham, *Third World Politics: An Introduction* (1985), p. 48. This definition is quoted *very* widely; see for examples, among many: Almezaini, "Private Sector Actors in the UAE and Their Role in the Process of Economic and Political Reform", *Business Politics in the Middle East*, ed. Hertog, Luciani, and Valeri (2013), p. 51; Mkandawire, "Neopatrimonialism and the Political Economy", p. 565.

[28] The points that follow are partly taken from: Bank and Richter, "Neopatrimonialism in the Middle East and North Africa: Overview, Critique and Alternative Conceptualization", presented at the GIGA workshop *Neopatrimonialism in Various World Regions* in Hamburg, Germany, 2010.

[29] Schlumberger, "Structural Reform, Economic Order, and Development: Patrimonial Capitalism", *Review of International Political Economy* 15.4 (2008), p. 626.

[30] Erdmann and Engel, "Neopatrimonialism Reconsidered: Critical Review and Elaboration of an Elusive Concept", *Commonwealth & Comparative Politics* 45.1 (2007), pp. 106–8.

[31] Schlumberger, "Structural Reform, Economic Order, and Development", p. 626.

[32] Heydemann, "Networks of Privilege: Rethinking the Politics of Economic Reform in the Middle East", *Networks of Privilege in the Middle East: The Politics of Economic Reform Revisited*, ed. Heydemann (2004), pp. 1–34; Sassoon, *Anatomy of Authoritarianism in the Arab Republics* (2016), p. 114.

[33] Herb, *All in the Family*, p. 15.

[34] Schlumberger, "Structural Reform, Economic Order, and Development", pp. 625–6.

[35] Almezaini, "Private Sector Actors in the UAE", pp. 50–9.

[36] Herb, *All in the Family*, p. 15.

societies are not as passive as often assumed. Nonetheless, societal forces can be politically marginalized when they are beneficiaries of handsome rentier allocations, but also because neopatrimonialism extends beyond institutions and into *social* groups, even societal forces, as well. People do not simply receive rents as individuals, but also through the groups and social forces of which they are part, such as tribes, key families, and civil society organizations.[37]

The Gulf's neopatrimonialism and its (late) rentierism are thus tightly linked. Since rent allocations are, as noted, very blunt instruments, neopatrimonialism hones allocative channels. Society may become accustomed to rent disbursements and see them as an entitlement; some observers argue for a "rentier mentality" where this is the case.[38] If so, neopatrimonialism provides more politically targeted and negotiable pathways for rent dispersal. One example is state employment. This is, in itself, a channel for rent allocation, but moreover, when someone takes up employment in a public-sector body or state-owned firm, they are entering into a new relationship with a personalized institution permeated by patron-client networks, which seeks to manage their (indirect) links to the state and condition them towards accepting the status quo. Not all rent travels through such mechanisms, but when neopatrimonial rent is combined with broader allocations, it is even more effective. Rents provide the funds that sustain patron-client networks and, through these, an open but non-democratic state-society communications channel. Neopatrimonialism, in turn, creates new avenues for rentierism to utilize.

Neopatrimonialism also helps to fill the political gaps that remain even in the most rentierist of systems. Even where rulers have tremendous allocative capacity, elites remain a potential threat. Some will simply be oppositionist and refuse to be co-opted, while others, such as clerics or merchants, have the ideological potential and the financial means, respectively, to oppose the regime. Rulers therefore usually make bargains with such elites, and seek to reach wider societal forces through them, allocating specific wealth, opportunities, and protection through such networks. Saudi Arabia is perhaps the most obvious case, where the Al Saʿūd have a co-optive bargain with key Najdī clerics, who are provided with rent-funded resources in exchange for their endorsement, and are consulted on occasional political matters of religious concern. The main merchant families, especially in Jeddah, are likewise co-opted. Similar mechanisms are in place in other Gulf states, even if the types of elites may vary.

Consistent with the personalization characteristic of neopatrimonialism, Gulf leaders also justify their rule in non-economic ways. One is through religion, and by undertaking to protect and preserve society's religious values. This has been the case very prominently in Saudi Arabia, where religion serves as an ideology of state formation.[39] Beyond religion, national narratives and historiography serve similar purposes in building a sense of shared nationhood and in positioning the state and royals in that collective imagination; again, Saudi Arabia provides a rich case study,[40] but it is a feature of most Gulf states.[41] Perhaps the most-discussed method of all in

---

[37] For a couple of more detailed examples, see: on social dynamics in the UAE, Rugh, *The Political Culture of Leadership in the United Arab Emirates* (2007), especially pp. 15–30 and 217–37; and on tribes and social forces in Qatar see: Crystal, "Tribes and Patronage Networks in Qatar", *Tribes and States in a Changing Middle East*, ed. Rabi (2016), pp. 37–55.

[38] The idea is widely argued in both scholarly and popular outlets, but was first outlined in Luciani, "Allocation vs. Production States", p. 88.

[39] Al-Dakhil, "Wahhabism as an Ideology of State Formation", *Religion and Politics in Saudi Arabia: Wahhabism and the State*, ed. Ayoob and Kosebalaban (2009), pp. 23–38.

[40] See the lengthy discussion of this in Al-Rasheed, *A History of Saudi Arabia*, pp. 182–210.

[41] Dessouki, "Social and Political Dimensions of the Historiography of the Arab Gulf", *Statecraft in the Middle East: Oil, Historical Memory, and Popular Culture*, ed. Davis and Gavrielides (1991), pp. 92–115.

recent times is national "branding",[42] in which most Gulf states have formulated a narrative of both national uniqueness and a more specific Gulf, or *khalījī*, identity. The former supports identity at a local and national level; the latter distinguishes Gulf identity from a wider and broader Arab one. Neopatrimonialism assists in this by demonstrating support for the state and its narrative among elites, by incorporating a non-economic dimension into the state-society relationship and by masking social divisions. In turn, such nation-building and state-building efforts strengthen the ruling elite and its rentier processes.

Finally, and very importantly for the Gulf's near- and medium-term future, is the potential for neopatrimonialism to serve as a stabilizing force during times of economic diversification and transformation, including occasions when reforms that are widely unpopular are nevertheless considered necessary. This is, as already noted, a fundamental vulnerability of rentiers: a "king's dilemma" occurs,[43] in which a ruler faces the need for change to economic or other reforms, yet lacks the elite flexibility or political inclusiveness to undertake such reform without either imposing it autocratically, and probably unsuccessfully, or attempting it and upsetting established elites and interests, thus risking losing power. A way around this "dilemma", however, is for reform to be negotiated at elite levels, and implemented through elite-sponsored, or "pacted", state-funded reform. While neopatrimonialism alone cannot act sufficiently to very broadly maintain public loyalty, it is very effective at solidifying elite loyalty to a ruler, thereby reducing what are perhaps the starkest threats to the ruler's survival.

That said, broad rent allocations are still required, but even changes to the allocative bargain can be negotiated through elite channels, as the planned 2018 introduction of a broad but low-rate consumption tax in the Gulf states demonstrates.[44] The late rentier idea assumes that regimes will seek to avoid unpopular policies, but it also argues that regimes, while not democratic, are still responsive, and that their overarching goal is long-term dynastic survival. Therefore, if they deem a certain reform unavoidable, they may adopt it slowly or cautiously, as with the consumption tax, but they will adopt it nonetheless. The rentier/neopatrimonial/state capitalist approach being outlined here in fact clarifies the introduction of unpopular reforms in a way that simple rentier theory did not: in short, if elite support for a reform can be negotiated, and if it seems likely it will outweigh the broad but uncoordinated opposition at societal level in general, then the reform will be adopted. Without elite agreement, leaders will instead have to build popular support for, or acquiescence to, the reform, which means doing so within the rentier framework instead — but this usually means counterbalancing the reform with trade-offs of some sort to society.

When the long-term development plans for the Gulf states are examined — the Qatar *National Vision 2030*, the UAE's *Vision 2021*, most recently and substantially the Saudi *Vision 2030*, and the others — they are remarkably similar in their strategy. They are not predicated on neoliberal reforms, but rather are focused on selective economic changes and with care paid to the social impacts of (rent-funded) reform.[45] As Hvidt notes, they also adopt a "cluster" approach, seeking to compete in selected clusters of similar or supporting industries, and take what he

---

[42] Cooke, *Tribal Modern: Branding New Nations in the Arab Gulf* (2014). For specific examples from Qatar, see: Kamrava, *Qatar: Small State, Big Politics* (2013); Gray, *Qatar: Politics and the Challenges of Development* (2013).

[43] Huntington, *Political Order in Changing Societies* (1968), pp. 177–91, cited and discussed in: Hudson, *Arab Politics: The Search for Legitimacy* (1977), pp. 166–8.

[44] Patchett-Joyce, "GCC's VAT Framework Takes Shape", *The National*, 7 May 2017.

[45] For more on these plans see: Hvidt, "Economic Diversification in GCC Countries: Past Record and Future Trends", *Kuwait Programme on Development, Governance and Globalisation in the Gulf States Research Paper* 27 (2013).

refers to as a "world-class city" approach.[46] To Hvidt, this suggests a high likelihood of failure, but their similarities also are a political manifestation of both the enduring rentier bargain and the importance of both neopatrimonial networks and new state capitalism structures to regime survival and legitimacy. What the Gulf states are proposing is not broad economic liberalization, which requires strong institutional autonomy and public accountability, but rather a more limited and politically safer strategy of selective economic opening, state-supported development, and a focus on sectors that do not threaten powerful incumbent elites and established interests. Since some social change will occur with such reforms, an elite agreement on reform is necessary, and is much more likely when it will occur through existing elites and channels, and will reinforce existing patterns of privilege.

Even the Saudi *Vision 2030* document, which goes considerably further than other Gulf strategies or earlier Saudi ones, will still be implemented in a rentier, state capitalist setting. It is substantial because it proposes amendment of some of the obligations of both state and society, strengthening non-oil economic sectors, embedding the economy more deeply in the global economy, and selling a small five-percent stake in the national oil company, Saudi Aramco, to free up capital to fund this diversification.[47] However, even if fully implemented, these strategies are partly extensions of earlier ones, especially those targeting economic diversification and investment. The the partial privatization of Saudi Aramco, while novel among the Gulf monarchies, would be small enough to pose no real threat to the state's control of the energy sector, and would have little impact on state revenue, yet at the same time, the Saudi government hopes that it will provide an enormous and immediate injection of some US$100 billion into socioeconomic development initiatives.[48]

4   Entrepreneurial state capitalism and rentierism

Late rentierism and neopatrimonialism make a formidable political partnership, but they are strengthened still further by the third dynamic in the triumvirate being explored here: "new", or "entrepreneurial", state capitalism. The concept of state capitalism is a well-established one, in the Middle East and elsewhere,[49] when, through nationalizations and greater state economic interventionism, it became a feature of many developing economies, including the Arab republics. A private sector and some market forces continued to operate, but the state became the dominant economic actor in these systems: it was the largest single owner of the means of production, and regularly intervened in the economy for political and social, not just commercial, purposes.[50] State capitalism also produced a state capitalist class that, while not *owners* of capital, continued a discrete group with shared political-economic interests that acted in the interests of the state and themselves.[51]

---

[46] Hvidt, "Planning for Development in the GCC States: A Content Analysis of Current Development Plans", *Journal of Arabian Studies* 2.2 (2012), pp. 189–207.

[47] For the details of the strategy, see the main public summary document on it, Govt. Saudi Arabia, *Vision 2030* (2016).

[48] The Saudi government has spoken of raising $100 billion with the sale of five percent of the company, which would value it at approximately $2 trillion. Some analysts place a lower value on it, varying from $500 billion up to $1.5 trillion, but few seem willing to accept the $2 trillion figure or to assume that the sale will proceed at that price. On the sale and some scenarios for it, see for example: Wald, "5 Ways a Saudi Aramco IPO Could Play Out", *Forbes*, 22 Oct. 2017.

[49] For some historical context on state capitalism see: Kurlantzick, *State Capitalism: How the Return of Statism is Transforming the World* (2016), pp. 49–63.

[50] On the case of Egypt see: Waterbury, *The Egypt of Nasser and Sadat: The Political Economy of Two Regimes* (1983), pp. 3–20; and on Syria see: Perthes, *The Political Economy of Syria under Asad* (1995), especially pp. 23–52.

[51] Waterbury, *The Egypt of Nasser and Sadat*, pp. 17–8.

The state capitalism of the contemporary Gulf states is different. It is not the product of revolutionary politics, nor driven by ambitions of social equity. Nor is it a variety of the state-led Asian "developmental state" models.[52] Even if some similarities exist, a range of differences also separates the two, including the fact that the Asian models are founded on an industrious indigenous working class and high rates of savings. Instead, Gulf state capitalism is a specific political-economic mechanism through which the ruling elite and its clients and supporters manage the economy, allocate and share wealth, and dominate neopatrimonial linkages. There is also good economic reasoning behind it, as Gulf rulers have long sought to avoid neoliberalism while also trying to diversify and develop the economy. While they are already the dominant owners and regulators in the economy, the case for a more entrepreneurial form of state capitalism was strengthened by the shock of low oil prices from the mid-1980s to the early 2000s.[53]

The Gulf's form of state capitalism is tagged as "entrepreneurial" here because it is a much more dynamic and risk-tolerant variety of state capitalism than that which characterized the republics in the 1950s and 1960s. It is what Bremmer calls a "new" state capitalism.[54] It is still "state capitalism", because the state is the dominant actor in the economy and the largest single owner of the means of production, but allows market-price mechanisms to operate and the private sector to play a (controlled) role. Yet it is "entrepreneurial", in the sense that state ownership has not been motivated simply by domestic social or political concerns.[55] States have sought to make these firms efficient, profitable, and innovative, using them not only to provide employment or subsidized goods, but also to diversify the economy and the sources of state revenue, carve out new economic niches, and promote the country internationally. The state still avoids politically risky neoliberal economic liberalization,[56] but also avoids the trap of *dirigisme* and the inefficiencies and political interests that it fosters or sustains.

This sets it apart quite starkly from the older state capitalism of the post-independence Arab republics where the state sought to dominate the planning process, ensure certain development outcomes for its popular social bases, develop heavy industry, and underwrite import substitution policies. In new state capitalism, in contrast, the state sets strategic goals and visions, rather than seeking to centrally plan or tightly manage the economy; favors the service sector and pursues "late-late-development" by bypassing most industrialization;[57] and encourages investment and export-driven growth, rather than pursuing import-substitution policies. It has several features. All the Gulf monarchies have state-owned but profit-driven and largely efficient national oil and/or gas companies,[58] as well as a strong sense of resource nationalism, in which hydrocarbons are treated as a strategic asset with both political and economic utility.[59]

---

[52] On the East Asian "developmental state" see: Pempel, "The Developmental Regime in a Changing World Economy", *The Developmental State*, ed. Woo-Cumings (1999), pp. 137–81.

[53] Maloney, "The Gulf's Renewed Oil Wealth: Getting it Right This Time?", *Survival* 50.6 (2008), p. 133.

[54] Of his several works on the topic and using this term, Bremmer's major piece is: *The End of the Free Market*. A number of other books have made similar points; for example, MacDonald and Lemco, *State Capitalism's Uncertain Future* (2015); and Kurlantzick, *State Capitalism*.

[55] This case is made in greater depth in: Gray, *Qatar*, pp. 10–1, 63–6, 67–70, 102–3.

[56] Bremmer, *The End of the Free Market*, especially pp. 51–81.

[57] Hvidt, "The Dubai Model: An Outline of Key Development-Process Elements in Dubai", *International Journal of Middle East Studies* 41.3 (2009), pp. 397, 398, 412–3.

[58] On this, see: Hertog, "Defying the Resource Curse: Explaining Successful State-Owned Enterprises in Rentier States", *World Politics* 62.2 (2010), pp. 261–301; and Hertog, "Lean and Mean: The New Breed of State-Owned Enterprises in the Gulf Monarchies", *Industrialization in the Gulf: A Socioeconomic Revolution*, ed. Seznec and Kirk (2011), pp. 17–29.

[59] Bremmer, *The End of the Free Market*, pp. 63–5.

Beyond oil and gas, there are powerful state-owned firms in sectors such as airlines, utilities, telecommunications, and others. As an example, most of Dubai's major firms are state owned, controlled either by the Dubai World or Dubai Holding groups,[60] while in Saudi Arabia firms such as Saudi Basic Industries Corporation (SABIC) and the National Commercial Bank may be publicly listed and have great autonomy, but are state controlled or majority state owned.[61] Other businesses and sectors — airlines, telecommunications firms, many financial services firms, and defense industries — are tightly controlled and often state owned. Like the national oil companies, these are typically efficient and profit-driven firms, providing the state with a stream of income (in most cases), or helping the country's international branding.

There are also private-sector firms that Bremmer calls "national champions",[62] which are favored by the state. Typically owned by royals or key merchant families, they sustain a business elite that is nearly always co-opted through neopatrimonial networks.[63] To thrive, most indigenous business needs either strong connections to, or the blessings of, a key member of the elite, and most foreign firms take a local partner to help with such connections. Even if informal, the state is in command of the economy and dominates the upper levels of the private sector.

Yet, even with these constraints, the Gulf monarchies are competing to create free trade zones, attract foreign direct investment (and increasingly portfolio investment as well), carve out niches in particular economic sectors, and reform their business rules and practices to make business processes simpler and cheaper. Thus, albeit conditionally, the Gulf's state capitalist structure is business and market friendly, but the state controls the largest and most valuable sectors of the economy. The state-owned firms have become a key venue in which neopatrimonialism is played out, with rulers able to dispense largesse and commercial opportunities through them, allot jobs to them,[64] or offer contracts to supply or partner with them. Therefore, as one example shows, key Qatari firms,[65] even when floated on the stock market, often remain under state control because the state will retain a critical number of shares or an underlying percentage of the firm that is not traded. Saudi Arabia provides a similar example, where some key firms are publicly traded, but with the state retaining majority control.[66]

Rentierism and entrepreneurial state capitalism relate to and reinforce each other in several ways. Most obviously, rents provide the funds that create and underwrite state-owned firms. Since new state capitalism primarily serves selected elites, rather than the population at large, wider co-optation through rentierism is an important, broader parallel to it. Indeed, a longer-term aim of entrepreneurial state capitalism — a key reason why it is *entrepreneurial* — is to provide alternative, stable, non-rent income for the state. Somewhat similarly, since the energy sector creates enormous wealth but not many jobs, insofar as entrepreneurial state capitalism helps diversify the economy it will offset the capital-intensiveness of the hydrocarbon sector. It also helps address other macroeconomic problems that stem from a reliance on rents, including

---

[60] Hvidt, "The Dubai Model: An Outline of Key Development-Process Elements in Dubai", p. 410.

[61] Luciani, "From Private Sector to National Bourgeoisie: Saudi Arabian Business", *Saudi Arabia in the Balance: Political Economy, Society, Foreign Affairs*, ed. Aarts and Nonneman (2005), p. 146.

[62] Bremmer, *The End of the Free Market*, pp. 67–9.

[63] Luciani, "From Private Sector to National Bourgeoisie", pp. 144–81 *passim*.

[64] Toksoz, "The GCC: Prospects and Risks in the New Oil Boom", *The Gulf Region: A New Hub of Global Financial Power*, ed. Nugée and Subacchi (2008), p. 89.

[65] For details see Gray, *Qatar*, pp. 67–70.

[66] This, for example, is the case with Saudi Arabian Basic Industries Corporation (SABIC), the country's main petrochemicals firm, which is publicly traded but is still 70% state owned, as well as with various other firms such as the Saudi Arabian Mining Company (Maʿaden), of which the state owns 50%, and the National Commercial Bank, among many others. See: Niblock and Malik, *The Political Economy of Saudi Arabia*, pp. 26–7, 136, 218–20.

economic bimodalism, where oil and gas draw investment and the best minds from the rest of the economy, creating two economies, a world-class energy one and a weak non-energy one, side by side. The state's strategy, therefore, is to use rentier income to invest in state capitalism, at least in part as a way to offset the risks and detriments of rentierism. If the Saudi sale of five percent of Saudi Aramco proceeds as planned, it will be an example of this: the sale of a very small share of the state's main rent-producing asset, but in exchange for an immediate injection of wealth to pay for a diversification that, it is hoped, will ultimately reduce the state's reliance on rent. The sale is only a minor amendment to the Saudi rentier-state capitalist system; if it works, it will consolidate this system, not reject or fundamentally reform it.

State capitalism is also a crucial component in the foreign policies of the Gulf states, both in their attempts to ensure protection from external threats and their attempts to build commercial and trade links with the rest of the world and, through the economic benefits that this brings, enhance their domestic support and legitimacy. These states, but especially the smaller shaikhdoms, conduct foreign policy with a strong set of economic goals, and often use state capitalism as a tool for this.[67] Foreign policy and state-owned firms work together to promote and coordinate national branding efforts, trade and investment policy, aid disbursements, and other economic policies.[68] Much of the attention given to sovereign wealth funds in recent years has pointed to their strategic and security ambitions.[69] They also serve a domestic political purpose, sending a message to society that the state is mature and engaged in long-term development thinking.[70]

5 Conclusion

The politics of the Gulf has been changing rapidly for the past two decades or more, and continues to change. In the process, the patterns of state and regime control are becoming more complex and sophisticated. Rents are still core to the political economy of the Gulf states and to the methods of rule deployed by Gulf leaderships. Yet, the older, simplistic variety of rentier theory is outdated. Rentierism only makes sense today when it is regarded, as is the concept of late rentierism, as a political dynamic and strategy. Its validity is strengthened still further, it has been argued here, when it is married with the concepts of neopatrimonialism and entrepreneurial state capitalism.

Rent, therefore, is not only a central mechanism through which the state co-opts and influences society, but, as this paper demonstrates, it links closely to these other two dynamics. Rents fund the main features and institutions of state capitalism, with neopatrimonial mechanisms used to manage the diffusion of rents and the reciprocal flow of loyalty and information back to leaders. In turn, neopatrimonialism provides a structure through which elite relationships can be managed by rulers — relationships that would otherwise be poorly addressed, or too bluntly managed if general rentier allocations alone were utilized. Entrepreneurial state capitalism supports rentierism by helping to compensate for its weaknesses and vulnerabilities, and by giving the state more choice and greater sophistication in the way it engages with both its population and the outside world.

---

[67] Ulrichsen, *Qatar and the Arab Spring* (2014), pp. 86–90.

[68] This argument is made in general terms in: MacDonald and Lemco, *State Capitalism's Uncertain Future*, pp. 101–29. On the case of Dubai, especially its use of soft power and economic power for security aims, see: Davidson, *After the Sheikhs: The Coming Collapse of the Gulf Monarchies* (2012), pp. 79–109; and some examples in: Krane, *Dubai: The Story of the World's Fastest City* (2009), especially pp. 151–95. On Qatar, see: Ulrichsen, *Qatar and the Arab Spring*, pp. 86–90; and Gray, *Qatar*, p. 210.

[69] Yi-chong, "The Political Economy of Sovereign Wealth Funds", *The Political Economy of Sovereign Wealth Funds*, ed. Yi-chong and Baghat (2010), pp. 7–16.

[70] Gray, "A Theory of 'Late Rentierism'", p. 34.

Finally, neopatrimonialism and entrepreneurial state capitalism also reinforce and support each other. The former is essential to how state capitalism operates and to ensuring that the political benefits derived from state capitalism reach the state and the ruling elite. State capitalism also provides a commercial realm in which the state can create and manage elites and develop the patron-client relationships that are such a salient feature of Gulf politics. Given the tight and reinforcing political webs created by these dynamics, Gulf regimes are extremely durable, and stasis is strongly favored by rulers and the political class. Only if state capacity can be markedly strengthened might the extant political order be fundamentally altered. Otherwise, as long as rulers and elites concur on the desirability of using rents for broad societal co-optation via more targeted channels, and elites willingly take up positions in neopatrimonial networks, these systems are unlikely to change — and if they do, change will only come from the top down.

Bibliography

1 Primary sources

Government of Saudi Arabia, *Vision 2030* (Riyadh: Saudi Government, 2016), available online at vision2030.gov.sa/download/file/fid/417.

International Monetary Fund (IMF), *Economic Diversification in Oil-Exporting Arab Countries* (Washington: International Monetary Fund, 2016), available online at www.imf.org/external/np/pp/eng/2016/042916.pdf.

2 Secondary sources

Al-Dakhil, Khalid S., "Wahhabism as an Ideology of State Formation", *Religion and Politics in Saudi Arabia: Wahhabism and the State*, edited by Mohammed Ayoob and Hasan Kosebalaban (Boulder: Lynne Rienner, 2009), pp. 23–38.

Almezaini, Khalid, "Private Sector Actors in the UAE and Their Role in the Process of Economic and Political Reform", *Business Politics in the Middle East*, edited by Steffen Hertog, Giacomo Luciani, and Marc Valeri (London: Hurst & Company, 2013), pp. 43–66.

Al-Rasheed, Madawi, *A History of Saudi Arabia*, 2nd edn (Cambridge: Cambridge University Press, 2010).

Al-Rasheed, Madawi and Robert Vitalis (eds), *Counter-Narratives: History, Contemporary Society, and Politics in Saudi Arabia and Yemen* (New York: Palgrave Macmillan, 2004).

Bank, André, and Thomas Richter, "Neopatrimonialism in the Middle East and North Africa: Overview, Critique and Alternative Conceptualization", paper presented at the GIGA workshop *Neopatrimonialism in Various World Regions*, Hamburg, Germany, 23 August 2010, available online at www.researchgate.net/publication/258325694_Neopatrimonialism_in_the_Middle_East_and_North_Africa_Overview_Critique_and_Alternative_Conceptualization.

Beblawi, Hazem and Giacomo Luciani (eds), *The Rentier State: Nation, State and the Integration of the Arab World* (London: Croom Helm, 1987).

Bremmer, Ian, *The End of the Free Market: Who Wins the War Between States and Corporations?* (New York: Portfolio, 2010).

Chaudhry, Kiren Aziz, *The Price of Wealth: Economies and Institutions in the Middle East* (Ithaca: Cornell University Press, 1997).

Clapham, Christopher, *Third World Politics: An Introduction* (Madison: University of Wisconsin Press, 1985).

Coates Ulrichsen, Kristian, *Qatar and the Arab Spring* (London: Hurst & Company, 2014).

Cooke, Miriam, *Tribal Modern: Branding New Nations in the Arab Gulf* (Berkeley: University of California Press, 2014).

Crystal, Jill, "Tribes and Patronage Networks in Qatar", *Tribes and States in a Changing Middle East*, edited by Uzi Rabi (Oxford: Oxford University Press, 2016), pp. 37–55.

————, *Oil and Politics in the Gulf: Rulers and Merchants in Kuwait and Qatar* (Cambridge: Cambridge University Press, 1995).

Davidson, Christopher M., *After the Sheikhs: The Coming Collapse of the Gulf Monarchies* (London: Hurst & Company, 2012).

Dessouki, Assem, "Social and Political Dimensions of the Historiography of the Arab Gulf", *Statecraft in the Middle East: Oil, Historical Memory, and Popular Culture*, edited by Eric Davis and Nicolas Gavrielides (Miami: Florida International University Press, 1991), pp. 92–115.

Eisenstadt, Shmuel, *Traditional Patrimonialism and Modern Neopatrimonialism* (London: Sage, 1973).

Erdmann, Gero and Ulf Engel, "Neopatrimonialism Reconsidered: Critical Review and Elaboration of an Elusive Concept", *Commonwealth & Comparative Politics* 45.1 (2007), pp. 95–119.

Foley, Sean, *The Arab Gulf States: Beyond Oil and Islam* (Boulder: Lynne Rienner, 2010).

Fox, John W.; Nada Mourtada-Sabbah; and Mohammed al-Mutawa (eds), *Globalization and the Gulf* (London: Routledge, 2006).

Gause, F. Gregory, *Oil Monarchies: Domestic and Security Challenges in the Arab Gulf States* (New York: Council on Foreign Relations, 1994).

Gengler, Justin, *Group Conflict and Political Mobilization in Bahrain and the Arab Gulf* (Bloomington: Indiana University Press, 2015).

Gray, Matthew, *Qatar: Politics and the Challenges of Development* (Boulder: Lynne Rienner, 2013).

————, "A Theory of 'Late Rentierism' in the Arab States of the Gulf", *Center for International and Regional Studies Occasional Paper* 7, Georgetown University School of Foreign Service, Qatar, 2011, available online at https://repository.library.georgetown.edu/bitstream/handle/10822/558291/CIRSOccasionalPaper7MatthewGray2011.pdf?sequence=5.

Herb, Michael, *The Wages of Oil: Parliaments and Economic Development in Kuwait and the UAE* (Ithaca: Cornell University Press, 2014).

————, *All in the Family: Absolutism, Revolution, and Democracy in the Middle Eastern Monarchies* (Albany: State University of New York Press, 1999).

Hertog, Steffen, "Lean and Mean: The New Breed of State-Owned Enterprises in the Gulf Monarchies", *Industrialization in the Gulf: A Socioeconomic Revolution*, edited by Jean-François Seznec and Mimi Kirk (London: Routledge, 2011), pp. 17–29.

————, "Defying the Resource Curse: Explaining Successful State-Owned Enterprises in Rentier States", *World Politics* 62.2 (2010), pp. 261–301.

————, *Princes, Brokers and Bureaucrats: Oil and the State in Saudi Arabia* (Ithaca: Cornell University Press, 2010).

Heydemann, Steven, "Networks of Privilege: Rethinking the Politics of Economic Reform in the Middle East", *Networks of Privilege in the Middle East: The Politics of Economic Reform Revisited*, edited by Steven Heydemann (Basingstoke: Palgrave Macmillan, 2004), pp. 1–34.

Hudson, Michael, *Arab Politics: The Search for Legitimacy* (New Haven: Yale University Press, 1977).

Huntington, Samuel, *Political Order in Changing Societies* (New Haven: Yale University Press, 1968).

Hvidt, Martin, "The Dubai Model: An Outline of Key Development-Process Elements in Dubai", *International Journal of Middle East Studies* 41.3 (2009), pp. 397–418.

————, "Planning for Development in the GCC States: A Content Analysis of Current Development Plans", *Journal of Arabian Studies* 2.2 (2012), pp. 189–207.

————, "Economic Diversification in GCC Countries: Past Record and Future Trends", *Kuwait Programme on Development, Governance and Globalisation in the Gulf States Research Paper* 27, The London School of Economics and Political Science, Department of Government (2013), available online at http://eprints.lse.ac.uk/55252/1/Hvidt%20final%20paper%2020.11.17_v0.2.pdf.

Jones, Toby Craig, *Desert Kingdom: How Oil and Water Forged Modern Saudi Arabia* (Cambridge: Harvard University Press, 2010).

Kamrava, Mehran, *Qatar: Small State, Big Politics* (Ithaca: Cornell University Press, 2013).

Krane, Jim, *Dubai: The Story of the World's Fastest City* (New York: St Martin's Press, 2009).

Kurlantzick, Joshua, *State Capitalism: How the Return of Statism is Transforming the World* (New York: Oxford University Press, 2016).

Luciani, Giacomo, "From Private Sector to National Bourgeoisie: Saudi Arabian Business", *Saudi Arabia in the Balance: Political Economy, Society, Foreign Affairs*, edited by Paul Aarts and Gerd Nonneman (London: Hurst, 2005), pp. 144–181.

————, "Allocation vs. Production States: A Theoretical Framework", *The Arab State*, edited by Giacomo Luciani (London: Routledge, 1990), pp. 65–84.

MacDonald, Scott B. and Jonathan Lemco, *State Capitalism's Uncertain Future* (Santa Barbara: Praeger, 2015).

Mahdavy, Hussein, "The Patterns and Problems of Economic Development in Rentier States: The Case of Iran", *Studies in Economic History of the Middle East*, edited by M.A. Cook (London: Oxford University Press, 1970), pp. 428–67.

Maloney, Suzanne, "The Gulf's Renewed Oil Wealth: Getting it Right This Time?", *Survival* 50.6 (2008), pp. 129–50, available online at www.brookings.edu/wp-content/uploads/2016/06/12_gulf_oil_maloney.pdf.

Marcel, Valérie and John V. Mitchell, *Oil Titans: National Oil Companies in the Middle East* (London: Royal Institute of International Affairs, 2006).

Mkandawire, Thandika, "Neopatrimonialism and the Political Economy of Economic Performance in Africa: Critical Reflections", *World Politics* 67.3 (2015), pp. 563–612.

Moore, Pete W., *Doing Business in the Middle East: Politics and Economic Crisis in Jordan and Kuwait* (Cambridge: Cambridge University Press, 2004).

Niblock, Tim and Monica Malik, *The Political Economy of Saudi Arabia* (London: Routledge, 2007).

Pempel, T.J., "The Developmental Regime in a Changing World Economy", *The Developmental State*, edited by Meredith Woo-Cumings (Ithaca: Cornell University Press, 1999), pp. 137–81.

Perthes, Volker, *The Political Economy of Syria under Asad* (London: I. B. Tauris, 1995).

Patchett-Joyce, Michael, "GCC's VAT Framework Takes Shape", *The National*, 7 May 2017, available online at www.thenational.ae/business/economy/gccs-vat-framework-takes-shape.

Ross, Michael L., *The Oil Curse: How Petroleum Wealth Shapes the Development of Nations* (Princeton: Princeton University Press, 2012).

Rugh, Andrea B., *The Political Culture of Leadership in the United Arab Emirates* (New York: Palgrave Macmillan, 2007).

Sassoon, Joseph, *Anatomy of Authoritarianism in the Arab Republics* (Cambridge: Cambridge University Press, 2016).

Schlumberger, Oliver, "Structural Reform, Economic Order, and Development: Patrimonial Capitalism", *Review of International Political Economy* 15.4 (2008), pp. 622–49.

Springborg, Robert, "GCC Countries as 'Rentier States' Revisited", *Middle East Journal* 67.2 (2013), pp. 301–9.

Toksoz, Mina, "The GCC: Prospects and Risks in the New Oil Boom", *The Gulf Region: A New Hub of Global Financial Power*, edited by John Nugée and Paola Subacchi (London: Royal Institute of International Affairs, 2008), pp. 81–96.

Valeri, Marc, *Oman: Politics and Society in the Qaboos State* (New York: Columbia University Press, 2009).

Wald, Ellen R., "5 Ways a Saudi Aramco IPO Could Play Out", *Forbes*, 22 October 2017, available online at www.forbes.com/sites/ellenrwald/2017/10/22/5-ways-a-saudi-aramco-ipo-could-play-out/#2dcb0bb038a2.

Waterbury, John, *The Egypt of Nasser and Sadat: The Political Economy of Two Regimes* (Princeton: Princeton University Press, 1983).

Weber, Max, *Economy and Society* (Tübingen: J.C.B. Mohr, 1922; reprinted in 2 vols by University of California Press, 2017).

Wehrey, Frederic M., *Sectarian Politics in the Gulf: From the Iraq War to the Arab Uprisings* (New York: Columbia University Press, 2014).

Yi-chong, Xu, "The Political Economy of Sovereign Wealth Funds", *The Political Economy of Sovereign Wealth Funds*, edited by Xu Yi-chong and Gawdat Baghat (Basingstoke: Palgrave Macmillan, 2010), pp. 1–25.

4   Reformers and the Rentier State

Re-Evaluating the Co-Optation Mechanism in Rentier State Theory

Jessie Moritz

Abstract: The oil and gas-rich states of the Gulf Cooperation Council have long been treated as exceptional, where distributions of rent-based wealth to society assumedly preclude political dissent. Yet, by examining informal and formal opposition in Qatar, Bahrain, and Oman since 2011, this article disputes the effectiveness of this "co-optation mechanism" at the sub-national level. Drawing from 135 semi-structured interviews conducted with citizens of these states, it uncovers evidence of challenges to state authority even among nationals who should theoretically be co-opted. In examining the limits of rent-based co-optation, the article highlights two key political dynamics that have demonstrated a capacity to overpower rent-based incentives to remain politically inactive: ideology and repression. Societies, then, were far from quiescent, and this research examines the networks and dynamics that have allowed citizens to challenge state authority.

1   Introduction

In 2011, the Arab world reeled from revolutionary unrest that toppled autocrats, fostered civil war, and generated widespread protests across much of the Middle East and North Africa. While scholars focused on Egypt and Tunisia puzzled over why they had failed to foresee such tumultuous change,[1] researchers working on the resource-rich states of the Gulf Cooperation Council (GCC) instead saw the relative *absence* of revolution in these states as vindication of the influential rentier state theory (RST). RST depicts petroleum-rich states as having an exceptional ability to resist democratization and popular accountability so long as resource wealth continues to flow. It is this understanding that prompted Michael Ross to ask in late 2011 whether oil would "drown" the Arab Spring, noting: "the Arab Spring has seriously threatened just one

---

Author's note: Financial support to conduct the research for this article was provided by the Australian National University as part of the author's PhD program. I would also like to thank the Center for International and Regional Studies (CIRS) at Georgetown University in Qatar for funding and organizing the workshop for which this paper was produced.

[1] Gause, "Why Middle East Studies Missed the Arab Spring: The Myth of Authoritarian Stability", *Foreign Affairs* 90.4 (2011), pp. 81–90.

oil-funded ruler — Libya's Muammar al-Qaddafi — and only because NATO's intervention prevented the rebels' certain defeat".[2]

Termed a "rentier bargain", the basic premise is that due to exceptional "rents", referring to the significant difference between cost of production and price on the international market for certain natural resources such as oil, the state can refrain from taxation, and distribute a portion of its wealth to society (where society primarily refers to citizens). As a result, incentives to challenge state authority are reduced in favor of rent-seeking for "rent distributions", in the form of material benefits, including loan forgiveness, direct cash transfers, subsidized utility rates, special land and loan packages, and public service employment. As leading RST theorist Giacomo Luciani declared in 1987: "the fact is that there is 'no representation without taxation' and there are no exceptions to this version of the rule".[3]

Much of the literature focuses on macro-level political outcomes, such as the incidence of revolution, democratization, or violent conflict, informed by a sub-trend within the early literature towards cross-national quantitative analyses. These works in turn rely on assumptions about political behavior at the micro- and meso-levels — after all, the exchange of material wealth for political legitimacy must work for at least a critical mass of individuals in order to produce these macro-level outcomes. Luciani states this quite explicitly when he reasons that, in a rentier economy, the incentive to individually petition the state for a greater share of rents "is always superior" to collective action challenging the system as a whole.[4] Although RST has developed considerably since then, scholarship on the political economy of the Gulf continues to defer to this framework, depicting the state as largely autonomous from a co-opted and passive society. Reform, where it does occur, is generally explained as the result of periods of reduced oil revenue, or due to top-down processes driven by the ruling elite.[5]

Within this context, the study of societal-driven reform movements has been sorely neglected. More problematically, the central causal link underpinning much of RST — that the distribution of rent-derived wealth necessarily leads to societal quiescence — remains largely unquestioned, especially at the sub-national level. By focusing on *exceptions* to the logic of the rentier state, this article reveals the limits of this "co-optation mechanism". It employs the congruence method of analysis, whereby the expected outcomes and causal mechanisms of RST are compared with actual outcomes and causal processes in three crucial real-world case studies: Qatar, Bahrain, and Oman.[6] Although all three fit the classic definition of a rentier state, in that rents formed at least 40% of government revenue at the start of the Arab Spring,[7] the selection of these cases was also designed to achieve a degree of concomitant variation, where the independent variable (rent-derived wealth) varies in expected directions with the dependent variable (societal quiescence), at least at the national level.[8] That is, Qatar represents the most rent abundant state (where rent abundance refers to rents per capita), and has experienced the least societal unrest; Oman represents a middling case, which experienced unrest in 2011 but has quietened since

---

[2] Ross, "Will Oil Drown the Arab Spring? Democracy and the Resource Curse", *Foreign Affairs* 90.5 (2011), p. 2.

[3] Luciani, "Allocation vs. Production States: A Theoretical Framework", *The Rentier State*, ed. Beblawi and Luciani (1987), p. 75.

[4] Ibid., p. 74.

[5] See, for example: Ehteshami and Wright (eds), *Reform in the Middle East Oil Monarchies* (2011).

[6] Eckstein, "Case Study and Theory in Political Science", *Case Study Method: Key Issues, Key Texts*, ed. Gomm, and Foster (2000).

[7] Kingdom of Bahrain, Economic Development Board, "Economic Yearbook 2013" (2013), p. 28; Sultanate of Oman, National Centre for Statistics and Information, "Statistical Year Book 2014", issue 42 (December 2014); Qatar Central Bank, "Thirty-Fourth Annual Report 2010" (2011).

[8] Peters, *Strategies for Comparative Research in Political Science* (2013), p. 30.

then; and Bahrain represents the least rent abundant case and has experienced widespread, intense, and ongoing protests since 2011.

To support this research, in-country fieldwork was conducted between June and July 2013 in the United Kingdom (to access opposition communities that had fled to the UK since 2011), and between September 2013 and February 2014 in the Gulf region. The intention was to supplement primary and secondary sources — such as government press releases; transcripts of public debates on state television and opposition-oriented satellite channels; opinion pieces in local newspapers; published lists of demands and manifestos of organized political societies; official Twitter accounts of members of government or prominent reformers; and country-specific online forums that functioned as a site of civic debate — with semi-structured interviews with citizens of Gulf states. The research resulted in 135 formal interviews — fifty-seven interviews in Oman, thirty-four in Qatar, and forty-four with Bahrainis based in Bahrain or the UK — designed to target those who were involved in reform movements since 2011, but also to capture a broad range of political views, including those who eschewed reform movements in favor of loyalist counter-demonstrations, and members of different factions within the state who hold diverse understandings of what the relationship with society is, or should be. Participants included: members of royal families; senior government bureaucrats, such as ministers and undersecretaries; elected and appointed representatives; economic and development advisors; members of the business elite; youth entrepreneurs; prominent leaders in civil society; members of political societies; and demonstrators involved in loyalist marches, opposition protests, or reformist movements since 2011.

Due to the level of trust needed to ensure interviewees were willing to share views on sensitive political topics, participants were identified using snowballing techniques, which by nature is vulnerable to selection bias.[9] In turn, this means only a fraction of politically active individuals could be analyzed in detail — in many cases, the more well-known figures. Yet, in authoritarian contexts, snowballing has been consistently identified as one of the most suited techniques, especially "when the focus of study is on a sensitive issue", and "requires the knowledge of insiders to locate people for study", a good example being members of the 14th February Youth Coalition in Bahrain, who are actively pursued by the state yet agreed to participate in this research.[10] Moreover, individual explanations for political action, particularly when carefully corroborated with group and national level data — for example published manifestos by political societies, and the reaction of mass demonstrations to government announcements of increased rent distributions — may highlight drivers of political mobilization that have been overlooked by RST, even if further research is needed to determine how and why sub-national dissatisfaction escalated, or failed to escalate, into mass political action.[11]

---

[9] While the findings of this research are described in as transparent a manner as possible, the balance between research transparency and the protection of interviewee identities is weighted heavily in favor of the latter, following advice in: Shih, "Research in Authoritarian Regimes: Transparency Tradeoffs and Solutions", *Newsletter of the American Political Science Association Organized Section for Qualitative and Multi-Method Research* 13.1 (2015), pp. 20–2. It is for this reason, for example, that the exact date of each interview cannot be published, and many identifying elements of interviewee identities have been omitted, particularly given the close-knit nature of Gulf societies.

[10] Biernacki and Waldorf, "Snowball Sampling: Problems and Techniques of Chain Referral Sampling", *Sociological Methods and Research* 10.2 (1981), p. 141.

[11] Existing work has identified several key factors that encourage political loyalty, or at least the ineffectiveness of national-level reform movements. See: Hertog, *Princes, Brokers, and Bureaucrats: Oil and the State in Saudi Arabia* (2010); Moore-Gilbert, "From Protected State to Protection Racket: Contextualising Divide and Rule in Bahrain", *Journal of Arabian Studies* 6.2 (2016), pp. 163–81.

The research particularly explores citizen justifications for why they personally engaged in "political action" since 2011, including participation in street demonstrations and in public debates through traditional media, in person, and through online forums such as Twitter, Facebook, WhatsApp, and country-specific websites. If RST is accurate at the sub-national level, should this not suggest that demonstrators were driven more by material factors (such as frustration with unemployment or dissatisfaction with economic benefits, indicating potential rent seeking) than by non-material motivations (such as desires for political liberalization or greater state accountability)? Politically active interviewees were accordingly asked to explain why, on a personal level, they had decided to engage in political action.

Notably, answers to this question varied markedly by political group: Omani youth protesters in Sohar, for example, outlined largely material motivations, such as frustration with continued unemployment, whereas leftist secular intellectuals in Bahrain claimed that their politicization occurred during interactions with pan-Arab nationalists abroad during university education in Lebanon, Syria, and Egypt. Identifying the underlying cause of political mobilization is, of course, extremely difficult. Protesters might claim in an interview that their mobilization was caused by a desire for greater political rights, whereas in reality they would not have mobilized had they not simultaneously been facing unemployment pressures. This research addressed this issue in several ways: by carefully evaluating the material and non-material drivers of political action expressed in in-depth interviews;[12] by examining the type of reform demanded (for example, better employment outcomes or greater political rights); and by documenting whether they ceased political action after receiving a rent distribution from the state, or, alternatively, suffered materially as a result of their political activity yet continued to demand political reform (the latter forming the most substantive challenge to the co-optation mechanism). This article is thus an investigation into the rentier bargain itself, but not solely from the perspective of the state (the most common approach, generally resting on the correlation between rent-funded benefits distributed by the state, and the relative absence of societal unrest at the national level), but also from the perspective of citizens themselves.

If non-material interests did drive societal reform movements, the question then becomes: why did these challenges emerge? What political, social, or economic variables have demonstrated the potential to overpower rentierism? Focusing on two of the dynamics most commonly referenced by opposition interviewees when asked to explain their personal motivation to press for reform, this article discusses how *ideology* and *repression* have overwhelmed rent-based incentives to remain loyal to the regime. Many citizens placed their privileged financial position in jeopardy by participating in public demonstrations, and the article discusses individual examples that challenge RST at the micro-level as well as broader macro-level opposition movements that highlight the continued relevance of societal activism in these archetypal rentier states.

2 A caveat: material drivers of dissent

This article focuses on non-material causes of societal reform movements, but this should not suggest that material factors were not also present. Indeed, the research completed for this project identified several material drivers of political action in the Gulf, including dissatisfaction with employment outcomes, anger at perceived state corruption, and frustration with

---

[12] "Non-material" interests and motivations, are defined here as "of political or social nature and not easily quantified or holding direct monetary value", whereas "material" interests (referring to basic material needs, such as those relating to bodily needs or wants) *could* be considered a form of rent-seeking.

"inadequate" government services; some of this material has been published elsewhere.[13] Most drivers fell within the scope of "inequality", referring to perceived discrepancies in income, employment, or other material benefits.[14] That is, rents have co-optive power, but *relative* deprivation and *relative* inequality can very quickly undermine rent-based co-optation, and, once mobilization occurs, activists may face repression that, in turn, solidifies their political opposition, which is essentially what occurred among Sohari youth protesters in 2011.

Beyond inequality, there was also evidence that rent distributions had ironically *contributed* to the emergence of reform demands, for example, by funding public education, which over time led to the uptake of ideologies or political cultures that encouraged political challenges.[15] Material and non-material drivers of societal reform movements, then, remain closely interlinked. Despite this, GCC states were eager to portray 2011 protests as related solely to economic rather than political grievances. Even King Hamad bin Isa al-Khalifa of Bahrain described the Bahraini Spring as "demands for well-paying jobs, transparency in economic affairs and access to better social services", and recommended entirely material responses to restore stability.[16] Theoretically, this would be considered a *failure* of the rentier state to effectively distribute wealth to society, not a challenge to RST itself. Inequality in rentier states is thus an important driver of political action, yet the almost exclusive focus on material motivations within RST also obscures other, non-material motivations for political action.

3   Rent-seeking and political challenges

RST focuses attention on the material basis of political legitimacy and authority, yet, as a result, neglects non-material dynamics shaping political activism. An emerging trend within RST works on the Gulf region, however, has started to interrogate core causal mechanisms of the theory, and, critically, they have started to do so at the sub-national level.[17] A survey of Bahraini citizens, for example, found that co-optation may be effective only for Bahraini *Sunni* respondents; for Bahraini Shiʿa respondents, the sense that the government was persecuting Shiʿa as a political group was enough to prompt even very wealthy Shiʿa — with little material cause to challenge the state — to join anti-regime demonstrations.[18]

These works represent an important attempt to move beyond an understanding of state-society relations as allocative-passive and towards something that can better explain the political activism of Gulf society, even in the absence of revolution or democratization. Yet they remain, for the moment, a minority within the broader literature. Differentiating between material and non-material causes of political dissent is, of course, exceedingly difficult. This article draws a critical distinction between *rent-seeking* and what is termed a *political challenge*. If "reformist" movements since 2011 are predominantly demands for greater rent allocations and benefits (for example, public sector employment) — even if this represents a justifiable attempt to improve

---

[13] Moritz, *Slick Operators: Revising Rentier State Theory for the Modern Arab States of the Gulf*, PhD diss. (2016).

[14] See also: Okruhlik, "Rentier Wealth, Unruly Law, and the Rise of Opposition: The Political Economy of Oil States", *Comparative Politics* 31.3 (1999), pp. 295–315.

[15] This finding is also corroborated by Martín, *Rentierism and Political Culture in the United Arab Emirates: The Case of UAEU Students*, PhD diss. (2014).

[16] Al Khalifa, "Al Khalifa: Stability Is Prerequisite for Progress", *The Washington Times*, 19 Apr. 2011.

[17] Coates Ulrichsen, *Qatar and the Arab Spring* (2014); Mitchell, *Beyond Allocation: The Politics of Legitimacy*, PhD diss. (2013).

[18] Gengler, *Group Conflict and Political Mobilization in Bahrain and the Arab Gulf: Rethinking the Rentier State* (2015).

one's material circumstances or address inequality — the fundamental logic of the rentier state remains upheld. If, however, citizens have made *political challenges* — demands to shift political power from state to society despite co-optation mechanisms being in place, such as a call for political liberalization by wealthy elites — then this signifies at least a partial rejection of the rentier bargain. Even a cursory glance at reformist movements in the Gulf, of course, reveals that they often embody elements of both. A protest decrying government corruption could be motivated by frustration that another actor has captured an inequitable share of rents. Addressing corruption, however, also implies greater government accountability — a political challenge. This was apparent in calls among demonstrators in Sohar for the removal of Ahmad bin Abdulnabi Makki, then Omani Minister of the National Economy, which were made both by activists that, upon being interviewed, expressed rent-seeking motivations, as well as among activists seeking political reform.

Even among organized political societies, separating material from non-material demands is exceedingly difficult. The Manama Document, released 12 October 2011, by five Bahraini opposition societies, justifies their activism almost entirely by reference to material dissatisfactions, including inequality in infrastructure, housing shortages, and overcrowding in Bahrain's only public hospital. The proposed *demands*, however, present a political challenge, including an elected unicameral parliament with legislative power, an independent judiciary, and the power to withdraw confidence in the Prime Minister and Cabinet "should they fail in their duties".[19] The document reveals the interplay between material and non-material demands; as Abdulnabi Alekry, a member of Bahrain's main secular leftist political society, Waad, noted, many Bahrainis view political reform as the "door" to reforming the broader system and tackling other problems "like a fair distribution of wealth".[20]

Citizens of GCC states have made both rent-seeking demands and political challenges of their governments since 2011, and, importantly, their justification for political mobilization may shift over time. Interviewees from the 14th February Youth Coalition in Bahrain, for example, expressed material motivations for joining demonstrations, such as dissatisfaction with employment outcomes. However, the politicized atmosphere of Bahrain and widespread use of repressive tactics by the state since 2011 also caused them to make clear political challenges (such as their call to "overthrow the regime" as "efforts to reform and coexist with the regime have become impossible").[21] Members of this group claimed in 2014 that they were unlikely to accept an entirely material response from the state, even if it personally improved their financial position.[22] Some had already lost employment, scholarships, or other material benefits, ostensibly because of participation in demonstrations; their continued activism following the loss of these benefits is difficult to explain within the context of RST, even if their initial politicization was driven by rent-seeking.

This article focuses on individuals and groups who were motivated primarily by non-material factors, and who made clear political challenges of the state. As noted above, the research did identify some groups that expressed material motivations for mobilization in 2011, and others that initially demonstrated but ceased their activism after receiving material benefits. They are not discussed here, since the surprising finding — in terms of identifying the limits of the co-optation mechanism — is not that rentierism remains relevant, but rather that the vast majority

[19] Bahrain Justice and Development Movement, "Manama Document", 13 Oct. 2011.

[20] Interview with Abdulnabi Alekry (A senior member of Waad), Bahrain, 2014.

[21] Anon., "I'tilāf shabāb thawra 14 Febrāyir" [Youth Coalition of the 14th February Revolution], *Maythāq al-lu'lu' lil-thowra al-rabi' 'ashar min Febriār* [Pearl Charter of the Youth Coalition of the 14th February Revolution], (2012), p. 3.

[22] Interviews (Members of the 14th February Youth Coalition), Bahrain, 2014.

of political actors examined for this research expressed non-material motivations to justify their political action, displayed little or no inclination to exchange material welfare for political quiescence, and remained politically active even after suffering the loss of material benefits.[23] This suggests that there exists at least a subset of individuals, even in these archetypal rentier states, for whom rent-based co-optation has been ineffective.

4   Rejection of the rentier bargain

Across the GCC, there exist individuals who campaigned for reform despite benefiting from the rentier state. Their activism is difficult to depict as rent-seeking. Many of these individuals disagreed *specifically* with the suggestion that citizens might exchange material welfare for political loyalty. "This is about dignity and freedom — it's not about filling our stomachs", said Ibrahim Sharif, the prominent Bahraini leader of Waad.[24]

Sharif's personal story reveals the weakness of materially focused explanations for Bahraini political loyalty. Born into a Huwala Sunni family in Muharraq in the late 1950s, he conforms to neither rentier nor sectarian depictions of the Bahraini Spring. His early experiments with politics occurred within the pan-Arab and anti-British atmosphere of the 1960s, which was especially influential in Muharraq.[25] When asked why he joined protests against the then-foreign-owned Bahrain Petroleum Company (BAPCO) in 1965 as a primary school student, he explained that: "You had no choice", citing the widespread politicization of society.[26] At university in Beirut in the mid-1970s, Sharif joined the Popular Front of Bahrain. When he returned to Bahrain in 1980, he was detained briefly by the Bahraini regime. "Really a turning point was Lebanon", he claimed in 2017. "Those were my 'forming' years and my politics became crystallized in activism .... The seed was there but there was nobody taking care of the seed and watering it. So it was in Lebanon where my interest in politics became an obsession".[27]

In 2002, within the context of King Hamad's reformist opening, Sharif helped to form Waad, eventually becoming its General Secretary. He ran for Majlis al-Nuwwāb elections in 2006 and 2010 and lost, yet demonstrated an ability to "galvanize the street in an unprecedented manner".[28] In 2011, Sharif quickly became a symbol for the cross-sectarian potential of Bahraini demonstrations.[29] Unsurprisingly, then, Sharif was arrested on 17 March 2011, and sentenced to five years' imprisonment for "conspiring to overthrow the government during street demonstrations".[30] State repression did not reduce Sharif's reformist aspirations. As he reported in 2017, having spent much of the previous six years incarcerated:

> If anything, that [repression] reinforced my desire, because I have now a personal experience of why we need the reform. Before that, it was not much on a personal level. It was an interest in the good of the society. But I was not personally — I was persecuted as part of the society, but not personally. But

---

[23] For a fuller overview of both rent-seeking and politically oriented activists, see: Moritz, *Slick Operators*.

[24] Fuller, "Bahrain's Promised Spending Fails to Quell Dissent", *The New York Times*, 6 Mar. 2011.

[25] AlShehabi, "Bahrain's Fate: On Ibrahim Sharif and the Misleadingly-Dubbed 'Arab Spring'", *Jacobin* 13 (2014).

[26] Interview with Ibrahim Sharif (Secretary General of Waad), Bahrain, 2017.

[27] Ibid.

[28] AlShehabi, "Bahrain's Fate".

[29] Sharif was usually the first example Bahraini opposition from all political groupings gave when asked their views on depictions of the protests as sectarian.

[30] El Gibaly and Jolly, "8 Bahrain Activists Get Life Sentences", *New York Times*, 22 June 2011.

this is an experience. And in experiences like this, you either end up — you're broken, and you end your political interest. Or you end up basically more determined.[31]

Sharif, then, came from an activist background and never displayed an inclination to exchange political loyalty for material welfare: rather, despite facing physical and material retribution for his political activism, he continued to demand reform, challenging RST's central logic.

Sharif might be considered an exception were there not a plethora of similar examples from across the GCC, indicating that there will always remain individuals who will challenge the state regardless of rent distributions. Other high profile examples include Qatari academics Dr Ali Khalifa al-Kuwari and Dr Hassan al-Said; Omani lawyer Basma al-Kiyumi; and former Majlis al-Dowla employee and writer Said Sultan al-Hashimi; all of whom have pressed for reform despite benefiting from the rentier system. Al-Kuwari, for example, has enjoyed a career in public sector academia and consulting on oil and gas development in Qatar (through the Qatar Gas Company and Qatar University, for example), even while advocating for transfer of power from state to society.[32] Unlike protesters who expressed rent-seeking motivations, the political activism of these individuals is difficult to convincingly connect to material concerns.

What, then, has encouraged these individuals and groups to challenge state authority? The sections below elaborate how *ideology* and *repression* encouraged citizens of the GCC states to challenge state authority. They are far from the *only* drivers of political action relevant to the demonstrations across the Gulf since 2011,[33] but they represent the two most prevalent and distinct patterns identified during field research in 2013 and 2014. Often, of course, reformers expressed more than one motivation for political action; there is thus a great deal of crossover between the thematic sections below.

5 Ideology

Gulf states have long been concerned with the potential for transnational ideological movements to foment domestic dissent; consider their repeated attempts to disempower Arab nationalist and other leftist groups during the 1950s and 1960s.[34] Ideology, as a driver of political mobilization, refers to a distinctive system of beliefs held by an individual or group that conditions their understanding of the political world, shapes relations between members of the group and outsiders, and can provide a moral/ideological imperative to challenge state policy.[35] Ideologies are reinforced by ideological networks, which support the "aggregation process from individual discontent to collective action", and by political entrepreneurs, who frame emotional, spiritual, or other grievances within an ideological structure, offering an ideologically grounded "solution" to perceived problems.[36]

Not all ideologies have equivalent effects on political mobilization.[37] Some ideologies may encourage loyalty rather than opposition to state authority. It would be reasonable to expect, for example, that followers of the Senior Council of Ulama (Saudi Arabia's highest religious

---

[31] Interview with Ibrahim Sharif (Secretary General of Waad), Bahrain, 2017.
[32] Al-Kuwari, "Curriculum Vitae: Dr. Ali Khalifa Al Kuwari", *Dr Al Kuwari* website (2018).
[33] For a more complete discussion, see: Moritz, *Slick Operators*.
[34] Crystal, *Oil and Politics in the Gulf: Rulers and Merchants in Kuwait and Qatar* (1990), pp. 81–3.
[35] Definition drawn in part from Costalli and Ruggeri, "Indignation, Ideologies, and Armed Mobilizations: Civil War in Italy, 1943–35", *International Security* 40.2 (2015), pp. 119–57; and Maynard, "Ideological Analysis", *Methods in Analytical Political Theory*, ed. Blau (2017).
[36] Costalli and Ruggeri, "Indignation, Ideologies, and Armed Mobilizations", p. 132.
[37] Ibid, p. 144.

body, appointed by the king) would be influenced by condemnation of anti-government protest as un-Islamic in March 2011.[38] By comparison, the ideology of opposition-oriented political societies has shaped their political activities since 2011. Members of Bahrain's largest (at least until its disbandment in 2016) political society, al-Wefaq, consistently averred that they saw their role as being a mainstream, moderate political society that formally engaged with the government.[39] They were initially reluctant, in early 2011, to join street demonstrations — a hesitation that damaged their credibility among Bahraini youth demonstrators. "They were dining with the King on February 14th", said an activist aligned with the 14th February Coalition.[40] Conversely, the use of radical tactics among opposition groups such as Haq, al-Wafa, and other movements that embraced a more combative set of beliefs about how best to pursue political change, frustrated the formal opposition. "You don't throw out dusty water unless you have clean water", argued a youth activist aligned with Waad, referring to the former groups' stated intention to overthrow the Al Khalifa monarchy without clarifying what political system they thought should replace it.[41]

Ideology thus matters to the nature and intensity of political mobilization. There were four broadly identifiable groups of political activists driven by ideology relevant to this research: a "reformist elite"; secular leftist nationalists; Islamists that subscribe to the "dominant" religion within a state (such as Salafis in Qatar, or Ibadis in Oman); and Islamists that subscribe to the "non-dominant" religion (such as Shi'a in Bahrain). Members of all these groups referred to ideology when justifying political challenges made in the post-2011 period, even as the nature of their ideology differs sharply, as does their relationship with the state itself.

6 Reformist elite

The term "reformist elite" captures a broad swathe of citizens who generally share the following attributes: highly educated or experienced, often identifying as intellectuals; members of prominent tribes, families, or otherwise connected to societal elites; their political challenges tend to focus on technocratic reform, even if the political changes required to meet their demands would be significant; and their writings or public statements suggest they view these reforms as both necessary for the political sustainability of their country, and normatively "right". Some have formed organized societies, such as the Omani Society for Writers and Literati in Oman (chaired by former Omani Ambassador to the United States, Sadiq Jawad Sulaiman) that, in practice, operates as a forum for reformers to discuss their views in an open and scholarly environment.[42] Others are prominent individuals, a good example being Emirati academic Abdulkhaleq Abdulla. Interviewees from this group explicitly rejected the term "opposition", instead describing themselves as "reformers" who recommend political change through engagement with the state.[43] Indeed, several had previously held senior government positions, and may thus view themselves as linked to the state itself

---

[38] Anon., "Hay'at kibār al-'ulamā' fī al-Sa'udiyya taḥarim al-muẓāhirāt fī al-bilād watiḥdir min al-ratbāṭāt al-fikriyya wal-ḥizbiyya al-munḥirfa" [Senior Body of Scholars in Saudi Arabia Prohibit Protests in the Country and Advise Caution against Deviant Ideological and Factional Affiliations], *Asharq al-awsat*, 7 Mar. 2011.

[39] Interviews (Members of Al Wefaq), Bahrain and the United Kingdom, 2013–14.

[40] Interview (Bahraini youth activist aligned with the 14th February Youth Coalition), United Kingdom, 2013.

[41] Interview (Bahraini youth activist aligned with Waad), United Kingdom, 2013.

[42] Interviews (Members of the Omani Society for Writers and Literati, OSWL), Oman, 2013.

[43] Interview with prominent Omani reformer Said Sultan al-Hashimi and other Qatari, Bahraini, and Omani reformers from the reformist elite, 2013–14.

(functioning almost as an outer layer of political elites), even if their political demands would entail a radical restructuring of the state-society relationship.

Qatari academic Ali Khalifa al-Kuwari provides a revealing example. Although there is no Qatari "opposition", in terms of an organized society that campaigns publicly against the state, political reform demands have emerged from private *majālis* (sg. majlis, referring to a forum or gathering held in the home). The most prominent of the majālis is a group of sixty Qataris headed by al-Kuwari, who in 2012 published the edited volume, *Al-sha'ab yurīd al-'iṣlāḥ fī Qaṭar'aydān* [The People Want Reform in Qatar Too]. His "Monday Meetings", held from 14 March 2011 to 6 February 2012,[44] inspired the volume, which included the following demands:

- Transparency in public finance and publicly owned assets, including the state budget;
- Transparency in major public policy decisions, including criticism of the opaque manner in which education, health, and constitutional reform have been implemented;
- Freedom of opinion and media;
- Greater separation between public and private interests and independence of public administration;
- Rectification of the population imbalance between expatriate and Qatari citizens, including specific reform of the Nationality Law of 2005;[45]
- Effective economic diversification to increase long-term economic sustainability;
- Transition to a democratic political system (in a specific, step-by-step, manner);
- The creation of a democratic GCC-wide union to enhance Qatar's security.[46]

The volume is, for Qatar, unusually direct in its call for wide-reaching political reform, representing a clear political challenge. Other participants in the meeting are also well-known reformers, such as Hassan al-Sayed, an associate professor of constitutional law at Qatar University and *al-Sharq* columnist who campaigns against the continued delay of elections for Majlis al-Shura. "There is no excuse for postponing the elections or extending the term of the current Shura Council", he argued in 2013. In earlier articles, he noted that decisions to extend the Advisory Council's term had previously been issued over ten times since 1982, yet "none of these decisions clarified what was the public interest they intended to serve".[47]

There is little material incentive for these individuals to challenge Qatari state policy; rather, when the book was released, al-Kuwari and members of his majlis faced a backlash from Qatari loyalists.[48] As is typical of the reformist elite, they advocated reform *within* the existing system, not as a direct challenge to it (although, as revealed above, they nevertheless challenged key elements of the rentier bargain).

7 Secular leftists

There is some crossover between Gulf reformist elites and another strand of intellectually oriented political actors: secular leftist/nationalist groups. Both ascribe to an internationalized intellectual tradition, and their political goals are often similar, in terms of advocating largely secular and

---

[44] Al-Kuwari, "Al-laqā' al-ithnayn" [The Monday Meeting], *Dr-Al Kuwari* website (2018).

[45] Al-Kuwari specifically claims the Nationality Law is unconstitutional and deprives Qataris of the "rights of citizenship". See: Al-Kuwari, "Qataris for Reform", *Al-sha'ab yurīd al-'iṣlāḥ fī Qaṭar 'aydān* [The People Want Reform in Qatar Too], ed. al-Kuwari (2012).

[46] Ibid.

[47] Al-Sayed, "Why Do We Fear an Elected Council?", *The Peninsula*, 13 June 2013.

[48] Interviews (Qatari reformers and loyalists), Qatar 2013–14.

gradual political reform. However, where reformist elites are often comprised of former members of government, or otherwise portray their reform demands as "non-political", leftist activists have a stronger history of political opposition — including violent opposition — to British imperialism and, following independence, the state. Although most have laid down arms since the Dhofar War ended in 1975, the ideology of resistance against state authority, and focus on defending "exploited" citizens, remains influential. As Sharif explained: "the perspective you gain from joining a leftist movement is not [just] the Pan-Arab part, it's the internationalism that you gain ... that we're part of a bigger international movement seeking justice and equality".[49]

In Bahrain many members of Waad were formerly members of the Popular Front or other Marxist-oriented groups; Waad represents to a large extent the vestiges of the pan-Arab nationalist movement in contemporary Bahrain. In the post-2011 period, members of Waad displayed little inclination to exchange material welfare for political quiescence. Asked if they thought the Bahraini demonstrations were motivated primarily by economic or political concerns, interviewees affiliated with Waad immediately rejected the former motivation. "We don't need money, we need dignity", said an independent activist formerly aligned with Waad.[50] "It wasn't about the economic [desires]", argued Abdulnabi Alekry, a senior member of Waad. "I think it is more complex of an issue. It is that the people's aspiration was — as the government had stipulated in the [National Action] Charter — for a real constitutional monarchy, but this was not fulfill [ed]. So it is not only economic. Economic, yes in theory, but it is [also] about freedoms, about the discrimination, about naturalisation".[51]

8   Islamists (dominant religion)

Other ideological movements are far from secular. Islamists that align with the "dominant" religion, referring to the religion followed by the ruling family or emphasized by the state (for example, Ibadism in Oman), are often loyalist in orientation, yet may challenge specific policies based on perceived threats to their religious interests. As Courtney Freer's study of Muslim Brotherhood movements in Qatar, Kuwait, and the UAE argues: "ideologically driven Islamist movements are even less likely [than ordinary citizens] to be placated by government payouts, making it more probable that they become powerful voices of political opposition in rentier states".[52]

This varies, of course, between the Gulf states, dependent on other political and social factors that affect the political influence of Islamist movements. In Oman, notable instances of Islamist opposition occurred in 1994 and 2005 — the former connected to the Muslim Brotherhood and the latter formed by conservative Ibadi adherents dedicated to spreading Ibadism — even as the contemporary Ibadi clerical class, in general, remains strongly loyalist.[53] In Saudi Arabia, too, the politicization of Saudi Salafis since the First Gulf War, has allowed Sunni Wahhabi Islamists to "initiat[e] a renegotiation of the social contract in Saudi Arabia", particularly since the Saudi rulers "cannot easily quash or oppose Islamist arguments, since they stake their right to rule on largely Islamic grounds".[54] Other conservative religious groups framed their reform demands around perceptions that the state is threatening their moral/religious interests by, for example,

---

[49] Interview with Ibrahim Sharif (Secretary General of Waad), Bahrain, 2017.
[50] Interview (Independent activist formerly aligned with Waad), UK, 2013.
[51] Interview with Abdulnabi Alekry (Senior member of Waad), Bahrain, 2014.
[52] Freer, *Rentier Islamism: The Influence of the Muslim Brotherhood in Gulf Monarchies* (2018), p. 2.
[53] Valeri, *Oman: Politics and Society in the Qaboos State* (2009), pp. 184–7.
[54] Okruhlik, "Making Conversation Permissible", *Islamic Activism: A Social Movement Theory Approach*, ed. Wiktorowicz (2004), pp. 263, 265.

allowing consumption of alcohol in public spaces: this was apparent in debate over alcohol sales and consumption among conservatives in both Bahrain and Oman since 2011.[55]

Although the ideology of these groups often encourages loyalty to the ruling elite — particularly where they perceive the state as protecting them against the belligerence of other religious groups, as is the case for many Sunnis in Bahrain, or Shiʿa in Kuwait — RST's ability to explain their political mobilization is limited, at best.[56]

9 Islamists (non-dominant)

Islamist groups who adhere to an interpretation of Islam that clashes with that of the state, by comparison, are more likely to align with the opposition. Their political activism may be further inflamed if they believe that their community is being deliberately persecuted by the state; consider, for example, the mobilization of Muslim Brotherhood activists in the UAE and Saudi Arabia.[57]

However, as previous research has already noted, "religion as the basis of political ideology should be carefully distinguished from religion as a marker of identity that may in turn be the basis of political action".[58] The Bahraini and Saudi states, for example, have long argued that Shiʿa political mobilization in the Gulf is driven by an ideological alignment with Iran, and specifically an intention to overthrow Sunni rulers and install new Shiʿi Islamic theocracies modelled on and beholden to Iran.[59] Yet, linkages between previously militant Shiʿa transnational groups and contemporary opposition groups do not necessarily denote ideological alignment with Iran, nor are Gulf Shiʿa groups necessarily motivated by a religious ideology.[60] As an example: one of Bahrain's most prominent activists, former president and co-founder of the Bahrain Centre for Human Rights (BCHR), Abdulhadi al-Khawaja, was part of a violent Iran-linked group, the Islamic Front for the Liberation of Bahrain (IFLB), in the 1980s. Yet, since his return to Bahrain, he has publicly eschewed political violence and intervention by foreign nations in the politics of Bahrain.[61] More importantly, other members of the BCHR, such as Nabeel Rajab, are secular, and the BCHR overall espouses secular rhetoric focused explicitly on human rights abuses and government accountability, not the type of revolutionary rhetoric or Islamist goals that characterized the IFLB.[62]

The demands of the BCHR are framed by nationalist concerns, using the international language of the human rights community. This is not unusual for opposition societies in the Gulf. As Coates Ulrichsen argues: "most Shiite organisations and parties in the GCC continued to regard the nation-state as their primary point of reference when articulating demands for reform. They thereby remained rooted in their domestic context and held a far more nuanced

---

[55] See for example: Trenwith, "Bahraini MPs Renew Push to Ban Alcohol", *Arabian Business*, 28 May 2014.

[56] Interviews with Bahrainis Sunnis from across the political spectrum, Bahrain, 2013–14; on Kuwait, see: Moritz, "The Easy Enemy: The Shia and Sectarianism in the Arab States of the Gulf and Yemen during the Arab Spring", *Middle East Minorities and the Arab Spring: Identity and Community in the Twenty-First Century*, ed. Parker and Nasrallah (2017), pp. 242–5.

[57] Freer, *Rentier Islamism*.

[58] Anon., "Ideological Mobilization in the Muslim World", *Policy Perspectives* 6.1 (2009), p. 79.

[59] Matthiesen, *Sectarian Gulf: Bahrain, Saudi Arabia, and the Arab Spring that Wasn't* (2013).

[60] Wehrey, *Sectarian Politics in the Gulf: From the Iraq War to the Arab Uprisings* (2014).

[61] Interviews (Members of Bahrain's human rights community affiliated with the BCHR), Bahrain and UK, 2013–14.

[62] See examples of IFLB rhetoric in: Alhasan, "The Role of Iran in the Failed Coup of 1981", *The Middle East Journal* 65.4 (2011), pp. 603–7.

attachment to trans-national loyalties than supposed by suspicious ruling elites".[63] Describing BCHR activities as driven by Iran, based on the former membership of a senior member in the IFLB, thus overlooks all of the BCHR's activities that *do not* fit the government narrative. This is not to suggest that the Shiʻa identity of BCHR members is irrelevant: to the contrary, it provides a strong collective identity, politicized community, and mobilizing rhetoric that supports organized political action, particularly, since, as the next section discusses, Shiʻa communities in Bahrain often perceive the state as deliberately persecuting them.

Overall, depicting the motivation or demands of any of these ideologically motivated actors as rent-seeking would be greatly reductive, even inaccurate. For these individuals, substantive political reform is necessary to redress the disparity in power between state and society. Rent distributions, then, cannot guarantee an absence of opposition, particularly not in states with a long and established history of political activism and where domestic and transnational ideologies have overwhelmed material incentives to remain quiescent.

10   Repression

Repression — the state's use of coercive force against political groups, including limitations to public space for opposition or other policies intended to constrain opposition-oriented political life — was the most common explanation for political mobilization among opposition interviewees. Yet the existing literature on how repression affects political mobilization is divided. Recent findings suggest that repression can *decrease* political mobilization (forced demobilization) in rent-rich states by greatly increasing the cost of political action, and the evidence for its effect on mass demonstrations is certainly compelling.[64] However, research for this article examined not only mass demonstrations, but also public expressions of dissatisfaction and demands for reform on social media, online forums, and opinion pages of local newspapers. Public statements are associated with lower levels of political risk than participation in street demonstrations (let alone violent political action, the focus of much of the literature on repression and political mobilization) and may explain why repression was identified here as such an important mobilizing, rather than demobilizing influence.[65]

A related literature on the impact of *emotions* on political mobilization, especially fear, indignation, and anger, may also hold relevance. Specifically, this article aligns with Costalli and Ruggeri, in suggesting that "fear", as the underlying driver of repressive demobilization, "can have divergent effects, depending on the power relation with the threatening actor. If the actor's power is overwhelming, fear may be dispiriting. But if the power of the menacing actor is comparable to the power of the threatened community, then escalation dynamics may take place".[66] This may explain why, in the context of the Arab Spring and optimism about the possibility for successful political change, so many citizens of Gulf states mobilized despite the very real threat of repression. Further, even when the actor's power is overwhelming, the continued activism of certain groups and individuals may be explained by a sense that the cost of ceasing political activism is higher than the repressive cost of political action — particularly for those whose family members remain incarcerated. This ultimately suggests that there is a dual effect

---

[63] Coates Ulrichsen, *Insecure Gulf: The End of Certainty and the Transition to the Post-Oil Era* (2015), p. 44.

[64] Girod, Stewart, and Walters, "Mass Protests and the Resource Curse: The Politics of Demobilization in Rentier Autocracies", *Conflict Management and Peace Science* 35.5 (2018).

[65] See, for example: Regan and Norton, "Greed, Grievance, and Mobilization in Civil Wars", *Journal of Conflict Resolution* 49.3 (2005), pp. 319–36.

[66] Costalli and Ruggeri, "Indignation, Ideologies, and Armed Mobilizations", p. 153.

of repression: it can simultaneously *disincentivize* political action by heightening its cost among some societal groups, while *incentivizing* mobilization among others, especially those who were personally targeted by state repression. Most importantly, it suggests that rentier states are in fact a great deal less exceptional than typically supposed; in many cases, they are just as susceptible to traditional drivers of political mobilization as their resource-poor counterparts. Thus, if RST is to retain relevance as an explanation of state-society relations, it must retain theoretical space for the impact of these non-material political dynamics.

This was particularly apparent in Bahrain, where repressive tactics have been most intensely utilized. Bahraini activist Ali Abdulemam, for example, traced his involvement in politics back to 1991, at age thirteen, when he saw friends and neighbors in Bahrain being badly treated or tortured. "I need justice", he said, claiming that being jailed for his activism only hardened his resolve.[67] "When you come from a Shiʿa family, you have a family member in jail", said a Bahraini civil rights activist, who linked his personal motivation for political mobilization to the imprisonment of his uncle.[68] Another reformer, now operating out of the US, traced his political activism to his father's arrest.[69] As a Bahraini citizen who had grown up in the UK but started campaigning for Bahraini civil and human rights straight out of university put it: "Everyone in Bahrain is political. Even breathing the air, the tear gas, is a political act". She linked her personal motivation for aligning with the opposition — in her case the 14th February Coalition — with the persecution of her extended family.[70]

There was also evidence that repressive tactics had solidified existing opposition, and could cause the transformation of rent-seeking into political opposition. As noted earlier, participants in the Sohar demonstrations in Oman described their initial motivation for joining demonstrations in 2011 as almost entirely rent-seeking, whereas they expressed a mix of political and rent-seeking demands during interviews in 2013 and 2014; they linked this change to government crackdowns on demonstrators and their personal experiences with repression.[71] The list of official demands they submitted to the government in February 2011 also reflects anger with repressive state responses: the first five of forty-three total demands relate specifically to ending repression and opening public space for peaceful demonstrations.[72] In Bahrain, a 14th February Coalition activist claimed he was strongly against any acts of violence — his example was a recent bomb attack involving police cars in Sitra — but that regime repression had legitimated the use of Molotov cocktails, fake bombs, and tire-burning. "Shaikh Isa Qasem announced that we could use Molotov cocktails to protect women and children and dignity ... protect our family", he explained, describing his participation in burning tires in the streets behind the Bahrain Formula One track as an attempt to ensure that international media coverage would have smoke in the background. "We want to show that things are wrong in Bahrain", he argued, "It's not about hurting people".[73] Whether or not the 14th February Coalition should be considered a peaceful movement is questionable, yet the key point is that repressive actions by the state have lent legitimacy to more radical opposition.

Another Bahraini claimed he usually considered himself politically neutral but after the crackdown on Pearl Roundabout on 17 February 2011, he felt he had to choose a position. "No doubt [I chose] the people, they were unarmed", he explained, while emphasizing he still didn't agree with

---

[67] Interview with Ali Abdulemam (Bahraini activist), UK, 2013.
[68] Interview (Bahraini civil rights activist), UK, 2013.
[69] Interview (US-based Bahraini reformer visiting the UK), UK, 2013.
[70] Interview (UK-based Bahraini reformer), UK, 2013.
[71] Interviews (Omani participants in Sohar protests), Oman, 2013.
[72] Al-Hashimi (ed.), *Al-rabīʿ al-ʿUmānī: qarāʾat fī al-sīyyāqāt wa al-dilalāt* [The Omani Spring: Readings into its contexts and meanings] (2013), pp. 365–7.
[73] Interview (Member of 14th February Coalition), Bahrain, 2014.

everything the opposition said and did. In Oman: "I don't see myself as a politician", said an Omani activist involved in organizing 2011 demonstrations, "I'm a humanitarian who has just been forced into politics".[74] Several other Omanis and Bahrainis who identified as human rights activists explained their involvement in 2011 demonstrations in similar terms, claiming repression by the state had resulted in their politicization.[75] Upon being asked for his reaction to his personal arrest and torture, prominent reformer Said Sultan al-Hashimi affirmed: "the price of speaking out just made me more determined to push for reform".[76]

In Bahrain and Oman, then, repressive tactics may have driven mass demonstrations underground, but on a personal level, they often solidified and in some cases radicalized opposition. This is particularly important considering the ongoing crackdown on opposition in Bahrain evident after mid-2014, since which time the government has dissolved both al-Wefaq and Waad, the main opposition societies open to engaging with the government; revoked the citizenship of high-profile Shiʿa cleric Isa Qasem; upheld a nine-year prison sentence against Ali Salman, the Secretary General of al-Wefaq; and indefinitely suspended Bahrain's only opposition-oriented newspaper, *al-Wasat*. In response, Bahraini protesters have taken to the streets, especially outside the home of Isa Qasem in Diraz and in other Shiʿa-majority villages. More worryingly, on 22 January 2017, Morteza al-Sanadi, the exiled leader of al-Wafa, one of Bahrain's more radical opposition movements, announced that they had turned to armed struggle, declaring that: "we have entered a new phase. We have one hand in the streets and the other on the trigger".[77]

In Qatar, the state has not used repression against society widely since 2011, although of course it also has not faced such intense public opposition as in Oman or Bahrain. Historical experience suggests the Qatari state is willing to use repressive tactics when opposition does emerge. A notable example of this was the 2005 announcement by Emir Hamad revoking the Qatari citizenship of over 5,000 members of the al-Ghafran clan of the al-Marrah tribe, ostensibly because they held dual citizenship with Saudi Arabia, but more likely due to the alleged role of several members in the 1996 counter-coup attempt.[78] The order was rescinded several months later, but, as of 2012, Amnesty International estimated that at least 100 individuals had yet to have their nationality returned.[79] Kamrava claims: "the state's efforts at marginalizing the al Marrah have ironically led to their increasing self-awareness and, at times, expressions of grievance against the state".[80] These fraught relations have also been exposed in the context of the continuing Gulf crisis; Mohammed bin Hamid bin Jallab al-Marrah was one of only two Qatari figures noted to have attended a much-vaunted Qatar "opposition" conference in London in September 2017.[81] Here too, then, repression can solidify, rather than reduce, opposition.

The memory of the state's response to supporters of the 1996 attempted coup contributes to a widespread perception among Qatari reformers that public demands for political reform could result in their termination from public employment, or other loss of benefits. "In order to go to the streets", noted one Qatari reformer, "you need to be ready to lose those things. And for a Qatari, there's really a lot to lose".[82] The costs of political activism in Qatar, then, are considerably higher than the perceived benefits, particularly when, as al-Kuwari found upon publication

[74] Interview (Omani human rights activist involved in 2011 protests), Oman, 2013.

[75] Interviews (Omani and Bahraini human rights activists), Oman and Bahrain, 2013–14.

[76] Interview with Said Sultan al-Hashimi (Omani reformer), Oman, 2013.

[77] Anon., "UN Shocked by Manama Regime's Execution of Three Activists", *Press TV*, 18 Jan. 2017.

[78] Kamrava, *Qatar: Small State, Big Politics* (2013), p. 111.

[79] Anon., "Annual Report: Qatar", *Amnesty International* website (2012).

[80] Kamrava, *Qatar*, p. 111.

[81] Anon., "Qatar: Opposition Conference", *Gulf States News* 1044, 21 Sept. 2017.

[82] Interview (Qatari reformer), Qatar, 2013.

of his book, loyalists are widespread in society. This suggests that the *combination* of the de-mobilizing influence of repression, coupled with phenomenally high rent abundance, and a state that is remarkably responsive to citizen concerns even in the absence of political liberalization,[83] can minimize expressions of dissent; even so, some Qataris may still challenge the state.

11 Conclusion

Despite the absence of national-level reform movements, citizens in Qatar, Bahrain, and Oman remain actively engaged and were in many cases willing to reshape the state-society relationship. These political mobilizations highlight the importance of examining alternative variables that can have a critical effect on political outcomes, overpowering rent-based incentives for political quiescence. To some extent, this is the result of a natural disconnect between theory and practice: even though RST is often treated as a holistic explanation of state-society relations, it cannot explain a polity in its entirety, nor should it be expected to.

The repeated occurrence of political challenges and conscious rejection of rent-based co-optation efforts nonetheless suggests that, if RST is to have continued relevance to the modern Arab states of the Gulf (or other rent-rich countries), it needs to allow greater theoretical space for an important political outcome — that is: *co-optation can fail*, especially when it coincides with more influential variables such as ideology or repression. Shifting the focus to examine where rentierism *does not* predominate, compared to the more general assumption that it is universally effective, generates interesting questions that will help to strengthen the literature's explanatory power, such as: in which circumstances does rent-seeking prevail, compared to when political challenges dominate the political scene?

The dynamics identified in this article were based on those referenced most commonly by interviewees; they are far from the last word on the subject but suggest that, at least in the case of those interviewed, these political variables can overpower rent-based co-optation. More research is needed to understand why some sub-national mobilizations escalated into mass unrest, while others remained small-scale protests; it would also be useful to examine how the *state* perceives the rentier bargain — why it at times prioritized non-material responses to societal unrest and other times increased rent distributions — and the impact this had on macro, meso, and micro-level demonstrations.[84] Nonetheless, the mobilization of Gulf society since 2011 serves as a critical reminder to political economists that materially focused theories offer only a partial explanation of societal politicization, and that unless both political *and* material dissatisfactions are addressed, similar mobilizations have the potential to occur in the future.

Bibliography

1 Primary sources

1.1 *Published primary sources*

Anonymous, Amnesty International, "Annual Report: Qatar" (2012), available online at www.amnesty.org/en/countries/middle-east-and-north-africa/qatar.

---

[83] Loyalist Qataris interviewed for this project tended to justify their support for the state by claiming that it was responsive to their material and non-material needs, and that the leadership was genuinely interested in addressing their concerns.

[84] Some of this has been done in Moritz, *Slick Operators*, but more systematic analysis is needed.

————, Bahrain Justice and Development Movement, "Manama Document", 13 October 2011, available online at https://web.archive.org/web/20111029181524/www.bahrainjdm.org/?p=901.

————, "I'tilāf shabāb thawra 14 Febrāyir" [Youth Coalition of the 14th February Revolution], *Maythāq al-lu'lu' lil-thowra al-rabi' 'ashar min Febrāyir* [Pearl Charter of the Youth Coalition of the 14th February revolution], 2012, available online at https://14f2011.com/files/MethaqAR.pdf.

————, Qatar Central Bank, *Thirty-Fourth Annual Report* 2010 (August 2011), available online at www.qcb.gov.qa/English/Publications/ReportsAndStatements/Pages/AnnualReports.aspx.

Kingdom of Bahrain, Bahrain Economic Development Board, "Economic Yearbook 2013", 2014, available online at https://docplayer.net/8512939-Kingdom-of-bahrain-economic-yearbook-2013.html.

Sultanate of Oman, National Centre for Statistics and Information, "Statistical Year Book 2014", issue 42, December 2014, available online at www.ncsi.gov.om/Elibrary/LibraryContentDoc/ben_Statistical%20Yearly%20Book%20%202014_04e9e6e5-868e-4d2a-bcf4-b1dfb2e0edd7.pdf.

1.2 *Interviews*

Abdulemam, Ali (Bahraini Activist), United Kingdom, 2013.
Alekry, Abdulnabi (Senior Member of Waad), Bahrain, 2014.
Al-Hashimi, Said Sultan (Omani Reformer), Oman, 2013.
Anonymous (Members of the Omani Society for Writers and Literati, OSWL), Oman, 2013.
————, (Activist aligned with the 14th February Youth Coalition), United Kingdom, 2013.
————, (Bahraini civil rights activist), United Kingdom, 2013.
————, (UK-based Bahraini reformer), United Kingdom, 2013.
————, (US-based Bahraini reformer visiting the UK), United Kingdom, 2013.
————, (Bahraini youth activist aligned with Waad), United Kingdom, 2013
————, (Independent activist formerly aligned with Waad), United Kingdom, 2013.
————, (Bahraini youth activist aligned with the 14th February Youth Coalition), United Kingdom, 2013
————, (Omani human rights activist involved in 2011 protests), Oman, 2013.
————, (Omani and Bahraini human rights activists), Oman and Bahrain, 2013–14.
————, (Members of Al Wefaq), Bahrain and the United Kingdom, 2013–14
————, (Members of Bahrain's human rights community affiliated with the BCHR), Bahrain and United Kingdom, 2013–14
————, (Members of the 14th February Youth Coalition), Bahrain, 2014
————, (Qatari reformers), Qatar, 2013–2014.
————, (Omani reformers), Oman, 2013–2014.
————, (Bahraini reformers), Bahrain and the United Kingdom, 2013–2014.
————, (Qatari loyalists), Qatar, 2013–14.
————, (Omani participants in Sohar protests), Oman, 2013.
Sharif, Ibrahim (General Secretary of Waad), Bahrain, 2017

2 Secondary sources

Alhasan, Hasan Tariq, "The Role of Iran in the Failed Coup of 1981", *The Middle East Journal* 65.4 (2011), pp. 603–7.

Al Khalifa, Hamad bin Isa bin Salman, "Al Khalifa: Stability is Prerequisite for Progress", *The Washington Times*, 19 April 2011, available online at www.washingtontimes.com/news/2011/apr/19/stability-is-prerequisite-for-progress.

Al Kuwari, Ali Khalifa, "Qataris for Reform", *Al-sha'ab yurīd al-'iṣlāḥ fī Qaṭar 'aydān* [The People Want Reform in Qatar Too], edited by Ali Khalifa Al Kuwari (Beirut: Al-Maaref Forum, 2012), available online at http://dr-alkuwari.net/sites/akak/files/qatarisforreform-translation.pdf.

————, "Curriculum Vitae: Dr. Ali Khalifa Al Kuwari", *Dr-Al Kuwari* website (2018), available online at http://dr-alkuwari.net/node/425.

————, "Al-laqā' al-ithnayn" [The Monday Meeting], *Dr-Al Kuwari* website (2018), available online at www.dr-alkuwari.net/mondaymeeting.

Al-Hashimi, Said Sultan (ed.), *Al-rabī'al-'Umānī: qarā'at fī al-sīyyāqāt wa al-dilalāt* [The Omani Spring: Readings into Its Contexts and Meanings] (Beirut: Dār al-Fārābī, 2013).

Al Sayed, Hassan, "Why Do We Fear an Elected Council?" *The Peninsula*, 13 June 2013, available online at www.thepeninsulaqatar.com/uploads/2016/08/10/pdf/ThePeninsulaJune132013.pdf.

AlShehabi, Omar, "Bahrain's Fate: On Ibrahim Sharif and the Misleadingly-Dubbed 'Arab Spring'", *Jacobin* 13 (2014), available online at www.jacobinmag.com/2014/01/bahrains-fate.

Anon., "Hay'at kibār al-'ulama' fī al-Sa'udiyya taḥarim al-muẓāhirāt fī al-bilād wa taḥḍar min al-ratbāṭāt al-fikriyya wa al- ḥizbiyya al-munḥirfa" [Senior Body of Scholars in Saudi Arabia Prohibit Protests in the Country and Advise Caution Against Deviant Ideological and Factional Affiliations], *Asharq al-awsat*, 7 March 2011, available online at http://archive.aawsat.com/details.asp?section=4&issueno=11787&article=611299&feature#.Vy3XtRV96Rs.

———, "Qatar: Opposition Conference", *Gulf States News* 1044, 21 September 2017.

———, "Ideological Mobilization in the Muslim World", *Policy Perspectives* 6.1 (2009), pp. 79–88.

———, "UN Shocked by Manama Regime's Execution of Three Activists", *Press TV*, 18 January 2017, available online at www.presstv.ir/Detail/2017/01/18/506700/UN-appalled-Bahrain-execution-three-Shia-activists-US-State-Department.

Biernacki, Patrick and Dan Waldorf, "Snowball Sampling: Problems and Techniques of Chain Referral Sampling", *Sociological Methods and Research* 10.2 (1981), pp. 141–63.

Coates Ulrichsen, Kristian, *Qatar and the Arab Spring* (London: Hurst & Co., 2014).

———, Insecure Gulf: The End of Certainty and the Transition to the Post-Oil Era (New York: Oxford University Press, 2015).

Costalli, Stefano and Andrea Ruggeri, "Indignation, Ideologies, and Armed Mobilizations: Civil War in Italy, 1943–35", *International Security* 40.2 (2015), pp. 119–57.

Crystal, Jill, *Oil and Politics in the Gulf: Rulers and Merchants in Kuwait and Qatar* (Cambridge: Cambridge University Press, 1990).

Eckstein, Harry, "Case Study and Theory in Political Science", *Case Study Method: Key Issues, Key Texts*, edited by Roger Gomm; Martyn Hammersley; and Peter Foster (London: Sage, 2000).

Ehteshami, Anoush and Steven Wright (eds), *Reform in the Middle East Oil Monarchies* (Reading: Ithaca Press, 2011).

Freer, Courtney, *Rentier Islamism: The Influence of the Muslim Brotherhood in Gulf Monarchies* (New York: Oxford University Press, 2018).

Fuller, Thomas, "Bahrain's Promised Spending Fails to Quell Dissent", *The New York Times*, 6 March 2011, available online at www.nytimes.com/2011/03/07/world/middleeast/07bahrain.html?_r=1&.

Gause, F. Gregory III, "Why Middle East Studies Missed the Arab Spring: The Myth of Authoritarian Stability", *Foreign Affairs* 90.4 (July/August 2011), pp. 81–90, available online at www. foreignaffairs.com/articles/middle-east/2011-07-01/why-middle-east-studies-missed-arab-spring.

Gengler, Justin, *Group Conflict and Political Mobilization in Bahrain and the Arab Gulf: Rethinking the Rentier State* (Bloomington: Indiana University Press, 2015).

El Gibaly, Lara and David Jolly, "8 Bahrain Activists Get Life Sentences", *New York Times*, 22 June 2011, available online at www.nytimes.com/2011/06/23/world/middleeast/23bahrain.html?_r=0.

Girod, Desha M.; Megan A. Stewart; and Meir R. Walters, "Mass Protests and the Resource Curse: The Politics of Demobilization in Rentier Autocracies", *Conflict Management and Peace Science* 35.5 (2018).

Hertog, Steffen, *Princes, Brokers, and Bureaucrats: Oil and the State in Saudi Arabia* (Ithaca: Cornell University Press, 2010).

Kamrava, Mehran, *Qatar: Small State, Big Politics* (Ithaca: Cornell University Press, 2013).

Luciani, Giacomo, "Allocation vs. Production States: A Theoretical Framework", *The Rentier State*, edited by Hazem Beblawi and Giacomo Luciani (London: Croom Helm, 1987).

Martín, Marta Saldaña, *Rentierism and Political Culture in the United Arab Emirates: The Case of UAEU Students*, PhD dissertation (University of Exeter, 2014).

Matthiesen, Toby, *Sectarian Gulf: Bahrain, Saudi Arabia, and the Arab Spring that Wasn't* (Stanford: Stanford University Press, 2013).

Maynard, Jonathan Leader, "Ideological Analysis", *Methods in Analytical Political Theory*, edited by Adrian Blau (Cambridge: Cambridge University Press, 2017).

Mitchell, Jocelyn Sage, *Beyond Allocation: The Politics of Legitimacy*, PhD dissertation (Georgetown University, 2013).

Moore-Gilbert, Kylie, "From Protected State to Protection Racket: Contextualising Divide and Rule in Bahrain", *Journal of Arabian Studies* 6.2 (2016) pp. 163–81.

Moritz, Jessie, "The Easy Enemy: The Shia and Sectarianism in the Arab States of the Gulf and Yemen during the Arab Spring", *Middle East Minorities and the Arab Spring: Identity and Community in the Twenty-First Century*, edited by K. Scott Parker and Tony E. Nasrallah (New Jersey: Gorgias Press, 2017).

————, *Slick Operators: Revising Rentier State Theory for the Modern Arab States of the Gulf*, PhD dissertation (Australian National University, 2016).

Okruhlik, Gwenn, "Making Conversation Permissible", *Islamic Activism: A Social Movement Theory Approach*, edited by Quintan Wiktorowicz (Bloomington: Indiana University Press, 2004).

————, "Rentier Wealth, Unruly Law, and the Rise of Opposition: The Political Economy of Oil States", *Comparative Politics* 31.3 (1999), pp. 295–315.

Peters, B. Guy, *Strategies for Comparative Research in Political Science* (Houndmills: Palgrave Macmillan, 2013).

Regan, Patrick M. and Daniel Norton, "Greed, Grievance, and Mobilization in Civil Wars", *Journal of Conflict Resolution* 49.3 (2005), pp. 319–36.

Ross, Michael, "Will Oil Drown the Arab Spring? Democracy and the Resource Curse", *Foreign Affairs* 90.5 (2011), pp. 2–7.

Shih, Victor, "Research in Authoritarian Regimes: Transparency Tradeoffs and Solutions", *Newsletter of the American Political Science Association Organized Section for Qualitative and Multi-Method Research* 13.1 (2015), pp. 20–2.

Trenwith, Courtney, "Bahraini MPs Renew Push to Ban Alcohol", *Arabian Business*, 28 May 2014, available online at http://m.arabianbusiness.com/bahraini-mps-renew-push-ban-alcohol-551845.html.

Valeri, Marc, *Oman: Politics and Society in the Qaboos State* (New York: Columbia University Press, 2009).

Wehrey, Frederic M., *Sectarian Politics in the Gulf: From the Iraq War to the Arab Uprisings* (New York: Columbia University Press, 2014).

5 Cursed No More?

The Resource Curse, Gender, and Labor Nationalization Policies in the GCC

Gail J. Buttorff, Nawra Al Lawati and Bozena C. Welborne

Abstract: Recent scholarship posits that the resource curse has gendered as well as economic effects on oil-rich economies, like those in the Middle East and North Africa, entrenching paternalistic relationships that disadvantage women's entry into the labor force. Upon closer examination, however, it appears that oil may not be the most compelling argument to explain Arab women's low presence in the workforce — especially since women's labor-force participation within the oil rich Gulf Cooperation Council (GCC) states is generally higher than the regional average. We contend that this is, in part, a byproduct of the countries' labor nationalization policies. Our analysis suggests that oil-driven development can in fact boost female labor force participation, revealing that rentierism as experienced in the GCC can actually have positive externalities for women.

1 Introduction

The resource curse, or "the paradox of plenty", has been used to explain myriad social, political, and economic misfortunes in countries with large hydrocarbon wealth, including, but not limited to, economic underdevelopment, a proclivity toward authoritarianism, and instability.[1] Within the resource curse literature, one of the more intriguing arguments introduced over the past decade connects oil wealth to women's status in society. In short, oil income discourages women from entering the labor force, which, in turn, marginalizes them both economically and politically.

Perhaps most famously, Michael Ross linked gender inequality to "Dutch Disease", arguing that gender inequality was a byproduct of the structure of the economy, with resource-based economies engendering particularly negative outcomes for women.[2] Ross ventured that a rise in oil

---

Author's note: The authors would like to thank Zahra Babar, Mehran Kamrava, Suzi Mirgani, the participants at the CIRS "Resource Curse in the Persian Gulf" Working Group meetings, and two anonymous referees for their helpful comments and valuable insights.

[1] The resource curse thesis was popular from the 1970s — an era of widespread oil nationalization — to the 1990s. However, newer work from scholars such as Jones Luong and Weinthal, *Oil is not a Curse: Ownership Structure and Institutions in Soviet Successor States* (2010) and Schake, "The Myth of the Resource Curse" (2012) calls into question the negative effects posited by the original thesis, instead citing such factors as weak institutions and governance as the culprits in lagging economic growth.

income affects female labor force participation in two ways. One is through the lack of growth in export-oriented factories, especially those in the textile and garment industry, that have historically brought women into the labor force. The other is through the distribution of oil rents, which reduces the need for a second income and thus the need for women to work outside the home. Liou and Musgrave also link oil to patriarchy, arguing that oil works indirectly to provide a regime the necessary rents and budgetary flexibility to sustain costly chauvinistic policies placating key conservative — often Islamist — interest groups.[3] Despite the importance given to the role of policies designed to restrict women's mobility and access, the authors provide little evidence of policies that would be sufficiently restrictive to inhibit women from participating in the labor force.[4]

Resource curse explanations are commonly used to rationalize often negative socioeconomic outcomes in the Middle East and North Africa (MENA), due to the region's disproportionate share of the world's oil and natural gas reserves. And the region does after all have one of the lowest rates of female labor force participation in the world.[5] Ross and Liou and Musgrave focus on MENA countries as particularly emblematic of this trend; and the systematic gender inequalities present in the Gulf Cooperation Council (GCC) states, in particular, are presumed to exemplify their conclusions. Overall, we see that the conventional academic wisdom has very discrete ideas about women's socioeconomic opportunities in the MENA, and paternalism is perceived as partially born of oil wealth, but rarely delves deeper into women's lived realities.

In contrast to recent work, we find that under certain conditions — as surprisingly exhibited in specific Middle Eastern states — oil can have a positive impact on one dimension of gender equality: female labor force participation. Our research reveals that women participate more in the workforce in the MENA countries that are abundant in oil — specifically in several GCC states — than originally expected in the extant resource curse literature. In fact, according to a recent report by the consulting group A.T. Kearney, "the GCC has seen significant progress in recent decades — female participation in the workforce has increased by 33% since 1993, and the GCC has been one of the regions that has achieved higher progress over the past two decades".[6] Almost all GCC countries — the exception being Saudi Arabia — have much higher rates of participation than the MENA regional average, and in some cases GCC female labor force participation rates even compare favorably with the West.[7]

To understand possible gendered effects of the resource curse, it is crucial to recognize that not all oil-based rentier states are created equal and, in fact, the distinct structures of their economies and rent distribution capacities likely impact women in different ways. In this paper, we focus on the GCC countries and reveal how the adoption of labor nationalization policies has had gendered effects on the citizen labor force in these states, shedding new light on the social

---

[2] Ross, "Oil, Islam, and Women", *American Political Science Review* 102.1 (2008), pp. 107–23; Ross, *The Oil Curse: How Petroleum Wealth Shapes the Development of Nations* (2012).

[3] Liou and Musgrave, "Oil, Autocratic Survival, and the Gendered Resource Curse: When Inefficient Policy is Politically Expedient", *International Studies Quarterly* 60.3 (2016), pp. 440–56.

[4] For example, one policy that they mention is mandating the head covering, yet only two countries in the Middle East formally require women to adhere to certain dress codes: Iran and Saudi Arabia. These dress codes do not directly impede women from joining the workforce and may well increase their access to public space. Finally, the authors ignore policies that have been enacted across the Middle East, including in both Saudi Arabia and Iran, to increase women's labor force participation.

[5] Shalaby, "The Paradox of Female Economic Participation in the Middle East and North Africa", *Baker Institute Issue Brief*, 3 July 2014.

[6] See: Willen et al., "Power Women in Arabia: Shaping the Path for Regional Gender Equality" (2016), p. 2.

[7] Buttorff, Welborne, and Al Lawati, "Measuring Female Labor Force Participation in the GCC", *Baker Institute Issue Brief* no. 18, Jan. 2018.

impact of the resource curse. We contend that labor nationalization policies, which stem directly from the GCC's oil wealth and relatively unique rentierism, are at least partially responsible for the increase in women's labor force participation in the GCC region. Of course, increased female labor force participation is due to multiple factors, including increased motivation of women to consider careers beyond the home and rising women's skill levels.[8] As we discuss below, key to the success of labor nationalization policies in increasing female labor force participation is women's high level of education, which makes them well suited to take advantage of the push to nationalize the labor force. Women's entrance into the labor force has been facilitated not only by increased education and by motivation but also, we argue here, through state policies that have not necessarily intended the inclusion of women as their ultimate goal.

Both labor nationalization and the expansion of access to education are policies financed by oil wealth, with the former policies facilitating women's initial entry into the professional workforce. In many respects, this is the reversed pathway from expected modes of gender empowerment elsewhere in the world. Typically, women first enter the labor force through low-wage, low-skill jobs — usually in agriculture and later in manufacturing — and only later accede to professional jobs. This makes the GCC a truly unique case where oil rentierism might actually be a blessing rather than curse for women, at least in terms of initial access to the labor market at a higher-skill level.

It is important to note that women's greater participation in the labor force has not necessarily translated into enhanced political representation, yet again confounding the conventional wisdom related to gender empowerment. It further problematizes the question of whether approaching gender parity in the workforce necessarily translates into enhanced female status in other realms; a question that also has important implications for tracking the substantive progress of women elsewhere in the world.

2 Labor nationalization and segmented feminization in the GCC

2.1 Background and motivation

In the last two decades, the GCC states have pursued policies to nationalize the labor force aimed at substituting native labor for the immigrant workforce that dominates key sectors, a peculiarity of the region. Every GCC state has pursued this tactic to some extent.[9] Essentially, as defined by Randeree, "these are schemes designed to reduce the extensive reliance on foreign labor by encouraging, often compelling, private sector industries to hire nationals instead of foreign expatriates".[10] Labor nationalization policies are not new; some date as far back as the 1970s.[11] In the 1980s, nationalization efforts focused on the public sector and state-owned

---

[8] Willen et al., "Power Women in Arabia: Shaping the Path for Regional Gender Equality".

[9] Willoughby, "A Quiet Revolution in the Making? The Replacement of Expatriate Labor through the Feminization of the Labor Force in the GCC Countries", working paper no. 2004-18, *American University* (2004); Rutledge et al., "Women, Labour Market Nationalization Policies and Human Resource Development in the Arab Gulf States", *Human Resource Development* 14.2 (2011), pp. 183–98; Hertog, "A Comparative Assessment of Labor Market Nationalization Policies in the GCC", *National Employment, Migration and Education in the GCC*, ed. Hertog (2012), pp. 100–2.

[10] Randeree, "Workforce Nationalization in the Gulf Cooperation Council States", *CIRS Occasional Paper* no. 9 (2012), p. 1.

[11] Hertog notes that, "Saudi Arabia was probably the first Gulf country to experiment with national employment rules" [Hertog, "Arab Gulf States: An Assessment of Nationalisation Policies", research paper no. 2014/1, *Gulf Labour Markets and Migration* (2014), p. 14].

enterprises, but beginning in the 1990s, the focus shifted to the private sector.[12] While most early nationalization efforts were poorly implemented, and thus largely ineffectual, second-generation policies have been "better thought out and more systematic in nature".[13]

Many of the second-generation policies have their beginnings in the mid-1990s, and continue to be updated and revised, with many only having borne fruit in the new millennium. Nationalization policies have taken on a variety of forms, from national vision statements to formal laws, quotas, the designation of certain occupations for nationals only, as well as financial incentives to bridge the gap between private and public sector wages. In most GCC states, labor nationalization often included indirect interventions, such as citizen-specific vocational training and human resource development aimed at equipping citizens with the necessary skills in demand within the private or public sectors. It is important to note that labor nationalization as a strategy is very specific to the GCC and is largely atypical of other rentier states. There are only a few other countries, such as Singapore and Malaysia, that have engaged in this type of economic policy in the modern era due to certain demographic concerns.[14]

Another unique feature of GCC rentier states is the extreme demographic imbalance in the labor force, which consists largely of non-citizen workers in both skilled and low-skilled occupations.[15] It is estimated that foreigners make up 60% or more of the labor force.[16] This demographic imbalance in both the population and labor force was the result of rapid oil-led state and economic development, which required both skilled and unskilled labor. Because the Gulf states, at the time, lacked sufficient labor supply, foreign workers from both the wider Arab world and Southeast Asia filled the void.[17] Labor market nationalization policies have thus emerged to redress these demographic imbalances as well as to address rising unemployment among GCC nationals.[18]

Even though hiring national labor has been standard for the public sector, nationalization has also been extended to the private sector. Despite generous national welfare provisions, even GCC governments worry about rising unemployment among youth and women. There is also a real need to substitute foreign labor with domestic labor due to a decrease in rents with the price of oil currently so low, and an inability to subsidize citizens' incomes directly to the same extent as in the past. Falling oil prices have resulted in major public spending cuts as well as a recent spate of government downsizing, most visibly in Saudi Arabia and the UAE.[19] Thus, a critical unwillingness to tax the national population coupled with an expensive cash transfer program

---

[12] Fasano and Goyal, "Emerging Strains in GCC Labor Markets", IMF working paper no. 4/71, (2004), p. 16.

[13] Forstenlechner and Rutledge, "Unemployment in the Gulf: Time to Update the 'Social Contract'", *Middle East Policy* 17.2 (2010), p. 40; Hertog, "Arab Gulf States", p. 7; Forstenlechner and Rutledge discuss first-generation policies adopted in the1990s during a period of low oil prices and second-generation ones. Hertog distinguishes between first and second-generation nationalization policies according to the content of policies: prescriptive, like quotas on national employment, vs. market incentives, like reducing the wage gap (p. 4).

[14] Ruppert, "Managing Foreign Labor in Singapore and Malaysia: Are there Lessons for GCC Countries?", World Bank Publication vol. 2053 (1999).

[15] According to Hertog, "all GCC labour markets share two fundamental outcomes: (i) private labour markets dominated by foreigners and (ii) the outsized role of government in the employment of nationals" [Hertog, "Arab Gulf States", p. 4].

[16] Randeree, "Workforce Nationalization", p. 3.

[17] Hertog, "Arab Gulf States", p. 5; Randeree, "Workforce Nationalization".

[18] Randeree, "Workforce Nationalization", p. 2; Anon., "Gulf Arab States may See Unemployment Rise, Labour Reforms Needed-IMF", *Reuters*, 20 Nov. 2013.

[19] Kerr, "Saudi Arabia Cuts Public Sector Bonuses in Oil Slump Fall Out", *The Financial Times*, 27 Sept. 2016; Anon., "Sack All Foreigners by 2020, Says Saudi Government", *The New Arab*, 10 May

in some countries, has resulted in the GCC embracing labor nationalization as a way to transfer funds to its citizens in a bid to appease them and address concerns of foreign labor dependence.[20]

The demographic imbalance creates a unique security concern, but also a "national identity" concern for the states of the Persian Gulf.[21] Most GCC states are attempting to reinforce a coherent national identity in the face of potentially restive citizen populations, especially in the aftermath of the Arab Spring. Outside of the economic reasons for adopting these policies, they can also tap into popular nationalist and xenophobic sentiments against a largely non-citizen labor force that is void of rights, but demographically in the majority. Per Willoughby, "there is a fear amongst the national populations that their political and cultural integrity is under siege".[22] Hiring professional women for jobs that used to be performed by non-citizen professionals has emerged as one way to address these fears in states where male citizens do not possess the requisite skill sets.

Labor nationalization strategies also allow the state to claim higher citizen employment, while recapturing income previously earned by foreign workers, which is largely remitted to their home countries. Willoughby emphasizes this in the case of Bahrain, where he claims feminizing the labor force "would presumably raise the living standards of the typical Bahraini household and reduce the fiscal pressure on the state".[23] For states such as Bahrain, Oman, and even Saudi Arabia, where the income-to-citizen population ratio is less favorable than in the extreme rentier states, feminization of the workforce could thus allow the state to recoup lost profits as well. Beyond security, there is also growing recognition of the other economic benefits of greater female participation in the labor force by society and the ruling establishment.[24]

2.2 The hidden opportunity of labor nationalization policies

Although all GCC states have adopted some sort of labor nationalization policy, the nature and extent of the policies and their implementation vary across the cases. Enforcement levels have also been variable, both across time — i.e., between first- and second-generation policies — and across countries. The "success" of a given labor nationalization initiative seems to correlate with the severity of the sociopolitical issues the policy is geared at addressing. Specifically, Bahrain, Oman, and Saudi Arabia are under greater pressure to adopt such policies, due to their larger citizen populations and higher rates of unemployment with less government revenue dedicated per citizen. These countries have thus been somewhat more forceful in their implementation of labor nationalization, especially Oman and Bahrain.[25]

Furthermore, GCC states run the gauntlet of mid- to high-income countries with varying levels of foreign labor penetration, which also differentiates their nationalization policies and creates different prospects for women in the labor force. Bahrain, Oman, and Saudi Arabia would be considered middle-income countries with private sectors hosting a sizable citizen

---

2017; Anon., "Nine Top Officials Sacked after Dubai Ruler's Spot Check Reveals Empty Desks", *The Telegraph*, 30 Aug. 2016.

[20] Willoughby, "A Quiet Revolution in the Making?", p. 10.

[21] The Gulf Labour Markets, Migration and Population (GLMM) Programme estimated that non-citizens comprised the majority of the population in Bahrain (54.87%, 2017), Kuwait (69.15%, 2015), Qatar (88.04%, 2015) and the UAE (88.68%, 2014) with only Oman (45%, 2016) and Saudi Arabia (36.83%, 2016) having non-nationals as a minority.

[22] Willoughby, "A Quiet Revolution in the Making?", p. 11.

[23] Ibid., p. 30.

[24] Willen et al., "Power Women in Arabia"; Felder and Vuollo, "Qatari Women in the Workforce", RAND working paper no. WR-612-Qatar (2008); Randeree, "Workforce Nationalization".

[25] Fasano and Goyal, "Emerging Strains in GCC Labor Markets", p. 4.

presence, while foreign labor dominates the private sector in Kuwait, Qatar, and the UAE. Steffen Hertog ventures that Oman and Bahrain's labor nationalization has progressed more rapidly due to their use of market incentives to make national employees more attractive to the private sector.[26] Yet, if exclusively market-based incentives were truly at work, the public and private sectors of the GCC would be dominated by women since they tend to outperform men in educational attainment and have considerably lower wage expectations. According to Hertog, women's "lower wage demands mean that, all other things being equal, it would be easier to transfer foreign-held jobs to them".[27]

With the GCC governments framing labor nationalization as a bid to reduce dependence on foreign labor, this frame provides an entry point into the labor market for a previously underrepresented group with untapped resources: female citizens. Importantly, without the already high levels of female enrollment in vocational training programs and universities, this shift in the workforce would not have happened.[28]

The mobilizing structure that allowed women to take advantage of labor nationalization policies in the 1990s and 2000s, expanded with access to higher education over the previous decades. In fact, Herb attributes the growth in women's labor force participation to improvements in women's education, which was made possible by oil.[29] More and more women entered Arab universities, especially during the 1990s, with their graduation rates further increasing in the 2000s. The majority of GCC countries now have university systems dominated by female graduates across disciplines. In fact, Qatar, Kuwait, and Bahrain have some of the highest female-to-male ratios in tertiary enrollment in the world, with ratios of 6.66, 2.24, and 2.18, respectively.[30] Moreover, women in the GCC not only outpace men in educational levels attained, but also in achievement.[31]

Lower rates of male enrollment in higher education and a shift toward information economics give women significant advantages over their male counterparts. Rutledge and Al Shamsi, for

---

[26] Hertog, "A Comparative Assessment", p. 6. It seems Oman and Bahrain have also made the greatest strides in labor nationalization policy due to their inability to dole out as much in oil rents and subsidies as their neighbors, who have smaller citizen populations and/or more substantial oil wealth, necessitating and incentivizing other ways to co-opt the citizenry.

[27] Ibid.

[28] In 2009, Kuwait, Qatar, the UAE, Oman, and Saudi Arabia each had a higher percentage — on average more than sixty percent — of women enrolled in university than the United States and the United Kingdom, which had lower women's enrollment (around 50%) at the time. Various explanations have been introduced to account for this educational shift from individual to state level. DeBoer and Kranov see the increase in overall university enrollment rates as a byproduct of the push to educate a broader swathe of the population [DeBoer and Kranov, "Key Factors in the Tertiary Educational Trajectories of Women in Engineering: Trends and Opportunities in Saudi Arabia, the GCC, and Comparative National Settings", *Science and Technology in the Gulf States*, ed. Siddiqi and Anadon (2017), p. 56]. Tsujigami, meanwhile, cites the importance of the Saudi *ulama* in green-lighting women's education and participation in the workforce in the 1990s [Tsujigami, "Higher Education and Changing Aspirations of Women in Saudi Arabia", *Higher Education Investment in the Arab States of the Gulf: Strategies for Excellence and Diversity*, ed. Eickelman and Abu Sharaf (2017), p. 42]. Ridge also offers insight into what may have motivated this trend in women's educational attainment, noting that women's education now has symbolic value in broader policy discussions related to human rights and development in the GCC [Ridge, *Education and the Reverse Gender Divide in the Gulf States: Embracing the Global, Ignoring the Local* (2014)]. Due to sex segregation in schools, Ridge also maintains that teaching is deemed an appropriate career choice for women. Education has opened up women's access to other jobs and the public sphere more broadly.

[29] Herb, *The Wages of Oil: Parliaments and Economic Development in Kuwait and the UAE* (2014), pp. 22–3.

[30] Anon., *The Global Gender Gap Report* (2015).

[31] Rutledge et al., "Women, Labour Market Nationalization Policies"; Herb, *The Wages of Oil*.

example, found that women were "more likely to have the skill-sets required and sought by the more productive [private] sectors of the economy".[32] They also found that women "dominate in certain vocationally oriented subjects".[33] In the UAE, Rutledge et al. note that 75% of those enrolled in information technology courses and 77% of those enrolled in science courses are women. While women in Saudi Arabia outnumber men in the fields of business and economics (1.2 to 1) and medicine and pharmacology (1.5 to 1).[34]

In addition to their education and skill sets, women are also poised to benefit from nationalization policies because of how some employers view work ethics across male and female nationals. One study, for example, found that employers tend to view Qatari women more favorably than Qatari men in terms of work ethic.[35] Another study found that female students in Qatar reported a higher interest in working in the private sector than did their male peers.[36]

The ability of these policies to substantively affect women's labor contribution and presence in the workforce may also be a byproduct of targeting so-called mid-level skill and wage industries where "nationals are most likely to be competitive on skills, and the expatriate-local wage gap is the narrowest".[37] Most women of working age in the GCC enter the labor force in the service sector — the public sector and education — not in low-skill or blue-collar occupations. For example, 53% of Qatari women worked in the education sector in 2004; no women worked in construction, trade, or manufacturing; and only six women were reported to be working in the tourism industry.[38] Similar trends in women's sectoral employment are observed across the GCC. The same levels of labor feminization do not exist in blue-collar jobs since they are typically associated with immigrant labor within the GCC.

Women are further advantaged in that most of the jobs targeted for labor nationalization are in the service sector and require specialized skills associated with higher education. In particular, the majority of GCC women are very competitive for public and private jobs at this level, since these jobs are usually clerical and administrative as well as technical, which is in line with educational trends for women in the region. If these labor nationalization policies had been adopted even a few decades earlier, there would not have been the necessary cadre of educated professional women to hire.

In fact, one of the most striking aspects of the inclusion of GCC women into the national workforce is how their path into joining it has diverged from the pathways previously encountered in North America and Europe, as well as the rest of the world. Prior research has documented women historically finding initial employment in more "proletarian" occupations, such as factory work or agriculture, before acceding to professional employment.[39] However, this process has been reversed in the case of the GCC, with women first entering professional and service-sector jobs. There is evidence that this reversed pathway is largely due to the adoption of labor nationalization policies — ones benefiting nationals of both genders — that happily coincide with a large pool of educated citizen women.

---

[32] Rutledge and Al Shamsi, "The Impact of Labor Nationalization Policies on Female Participation Rates in the Arab Gulf", *Women, Work and Welfare in the Middle East: The Role of Socio-demographics, Entrepreneurship and Public Policies*, ed. Chamlou and Karshenas (2016), p. 527.

[33] Ibid., p. 535.

[34] Rutledge et al., "Women, Labour Market Nationalization Policies", p. 187.

[35] Felder and Vuollo, "Qatari Women in the Workforce", pp. 24–5; Stasz; Eide; and Martorell, *Post-Secondary Education in Qatar: Employer Demand, Student Choice, and Options for Policy* (2007).

[36] Felder and Vuollo, "Qatari Women in the Workforce", p. 28.

[37] Hertog, "A Comparative Assessment of Labor Market Nationalization Policies in the GCC", p. 33.

[38] Felder and Vuollo, "Qatari Women in the Workforce", p. 15.

[39] Willoughby, "A Quiet Revolution in the Making?", p. 34.

Furthermore, these social trends coincided with the growing importance of women in conveying a modern image of the Middle East to the world at large. In recent years, there have also been more overt attempts by GCC governments to formally include women into the labor force through gender quotas and by engaging in gender mainstreaming practices, albeit at a low level across public industries.

Extant research investigating how individual GCC women perceive these socioeconomic shifts has many of them opining it is a positive development, even though educated women are still more likely to be unemployed or underemployed than their male counterparts.[40] In 2013, Fauzia Jabeen surveyed some 200 students regarding their experiences with nationalization policies in the UAE, popularly known as "Emiratization". Seventy-eight percent of the women interviewed felt the process had actually helped them in their careers, substantially more than men, and 94% of women also saw education as key to their career progress. Jabeen notes that "they realise Emiratisation will only work for them if they're qualified", since the women compete against other Emiratis.[41] Zerovec and Bontenbal investigated Omanization policies promoted in 2003 that were aimed at increasing women's labor force participation from 6% to 12%, and found them to be successful. The authors note, however, that the private sector would have to "become more attuned to Omani women's needs and job aspirations", including offering higher salaries and childcare, in order to incentivize more women to enter the sector.[42]

The aforementioned nationalization policies in the UAE and Oman integrate women into the labor force without instituting any kind of targeting programs or selective policies that would otherwise make it look like they received preferential treatment for their gender. Rather, these women were offered jobs based on their nationality — a practice that could ultimately help bypass some of the stigma associated with gendered appointments or quotas, since the latter are often understood to imply that women do not have authentic qualifications.[43] In this way, labor nationalization policies have been used to circumvent cultural barriers, and as a way to reduce stigma associated with women working in the private sector. In addition, labor nationalization efforts geared at redressing the demographic imbalance between citizens and non-citizens in the workforce endow women with strategic importance. According to Randeree, "evidence clearly suggests that for prospective knowledge-based economies in the GCC region to succeed, they must first utilize all of their human resources [i.e., women] and encourage increased participation of women in the workplace".[44]

Ultimately, as noted by Randeree, "the pressure to increase the participation of nationals in the workforce has improved the prospects of employment for women in GCC states".[45] Although nationalization policies have not been explicitly gendered policies, they have, as we show here, gendered outcomes; and there is growing recognition that further nationalization efforts need to be gender-aware.[46] According to Rutledge et al., "it was noteworthy that the majority of the policymakers considered this to be a shortcoming that needed addressing. Not least

---

[40] Rutledge et al., "Women, Labour Market Nationalization Policies".

[41] Jabeen quoted in Swan, "Emiratisation Brings Greater Equality for UAE Women", *The National*, 29 Apr. 2013.

[42] Zerovec and Bontenbal, "Labor Nationalization Policies in Oman: Implications for Omani and Migrant Women Workers", *Asian and Pacific Migration Journal* 20.3–4 (2011), p. 378.

[43] Ultimately, this could have far-reaching implications for women who run for office and their ability to garner authentic popular support.

[44] Randeree, "Workforce Nationalization", p. 24.

[45] Ibid., 5.

[46] Rutledge et al., "Women, Labour Market Nationalization Policies", p. 192.

because many of the newly graduating national women were considered ... to be the most suitable candidates for the 'skilled' private sector".[47]

3 Labor nationalization policies and their gendered outcomes

This section delves more explicitly into how changes in labor policy incentivized women's entry into the workforce. There are several ways to analyze the impact of labor nationalization policies on women's labor force participation. The first and most straightforward way is to examine national-level labor force participation rates. Figure 1 presents female labor force participation as a percent of the female population, as opposed to the percent of the total labor force. The large non-citizen labor pool in the GCC, which also skews heavily male, can obscure the real participation rate of female nationals, likely underestimating it.[48] Somewhat inconsistent government statistics suggest labor force participation rates of citizen women are lower than those depicted in Figure 1, but higher than if total labor force is used as the denominator in calculating participation.[49] For example, according to the Qatari statistical authority, there were 166,013 Qataris aged fifteen years and older in 2012. Of those, 83,230 were female; of which 26,992 or approximately 32% were employed.

From Figure 1, we can see that female labor force participation rates steadily increased between 1990 and 2014, which corresponds to first- and second-generation labor nationalization efforts. Oman, in particular, saw its female workforce exceed the MENA regional average beginning in the late 1990s, after which it continued to climb. This tracks with the country's initial Omanization policy in 1994,[50] and its subsequent broadening in the first decade of the twenty-first century.[51] Similarly, in the Emirate of Abu Dhabi, government statistics show that the female labor force participation as a percent of employed

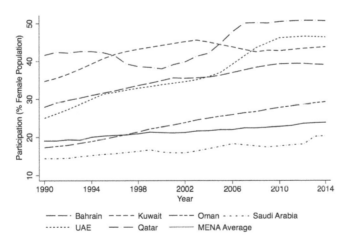

Figure 1: Female Labor Force Participation in the GCC, 1990–2014.
Source: World Bank, 1990–2014.

---

[47] Ibid., p. 192.

[48] This is a surprisingly consistent flaw in much extant analysis of women's labor trends in the region — i.e. discounting the distinct demographics of the GCC.

[49] Government statistics are not consistently available in similar form for every year for all countries.

[50] Fasano and Goyal, "Emerging Strains in GCC Labor Markets".

[51] Hertog, "An Assessment of Nationalisation Policies", p. 8.

citizens — i.e., excluding non-citizens — jumped from 10.47 in 1995 to 18.50 in 2001 and to 21.23 in 2008.[52]

Figure 1 also highlights that the GCC countries, with the exception of Saudi Arabia, have rates of female labor force participation higher than the MENA average for most of the period. In fact, the GCC average is close to the global average, ranging from approximately 30% to 40%.[53] This is particularly important because the MENA region as a whole continues to report the lowest female labor force participation rates in the world (approximately 25%), and one of the largest gender wage gaps (estimated to be between 20% and 40%).[54] Given that in many of the GCC countries women are participating at such rates, oil may not be the most compelling argument to explain Arab women's low presence in the regional work-force overall, as Ross and others posit. Rather, oil may actually have a positive gendered impact under certain conditions, facilitating, not disadvantaging, citizen women's entry into the labor force.

3.1 *Sectoral distribution of women in the labor force*

A second way to examine the gendered impact of labor market nationalization policies is to investigate the distribution of nationals across different economic sectors (e.g., public vs. private) and economic activity (e.g., finance, construction, etc.). In this paper we focus more on the financial services sector, primarily because it has consistently been singled out for nationalization quotas. For example, Oman set a quota of 45% for the Omanization of finance and insurance services, as well as for the real estate market. Kuwait set an initial target of 50% for banks, which increased to 66% in 2014.[55] Again, it is important to note that these quotas were gender neutral, but were not as predominant in other sectors or industries. Moreover, as noted previously, there has been a lot of inconsistency and variation in how nationalization targets have been enforced, but the financial services sector stands out for the importance national governments have attached to it. In the ensuing section we explore individual GCC cases, providing a brief overview of labor nationalization efforts and their impacts on women. For ease of readability, we discuss each country in turn.

3.1.1 *Bahrain*

Bahrain introduced a quota-based system in 1995 as part of its nationalization efforts, even though it already had one of the highest proportions of citizens employed within its private sector compared to other GCC states at the time.[56] Every new company was required to ensure at least 20% of its employees were citizens by the end of the first year of operation. The policy further required companies to increase national employees by 5% every subsequent year until a 50% nationalization target was achieved.[57] Table I reflects how Bahrainization policies helped shape higher

---

[52] Abu Dhabi Govt, "Demographic & Social Indicators", *Abu Dhabi Statistics Centre* (2008).

[53] Anon., *The Global Gender Gap Report* (2015).

[54] Ibid.; Shalaby, "The Paradox of Female Economic Participation"; World Bank, "Gender and Development in the Middle East and North Africa: Women in the Public Sphere" (2004).

[55] Randeree, "Workforce Nationalization"; Forstenlechner and Rutledge, "The GCC's 'Demographic Imbalance': Perceptions, Realities and Policy Options", *Middle East Policy* 18.4 (2011), pp. 25–43; al-Majdali, "National Labor Quotas Increased in Private Sector", *Kuwait Times*, 8 Aug. 2014.

[56] Hertog, "Arab Gulf States", p. 4. According to Hertog, Bahrain, Oman, and Saudi Arabia have the highest proportion of nationals employed in the private sector.

[57] Hertog, "Arab Gulf States", p. 1.

Table I: Distribution of Employed Population in the Public and Private Sector by Nationality and Gender, Bahrain, 2016.

	All (Nationals and Foreigners)		Nationals	
	Public Sector	Private Sector	Public sector	Private Sector
Men	35,813	512,105	29,138	72,398
Women	30,392	65,550	27,036	31,139
Total	66,205	577,655	56,174	103,537
Percent Women	45.9%	11.3%	48%	30%

Source: Information gathered from Government of Bahrain, "Table A1: Estimated Total Employment by Citizenship and Sector (Male) by Citizenship and Sector: 2006–2017" (2017).

female labor participation among Bahraini nationals compared to female participation rates among the population overall. Similar to other GCC states discussed below, aggregate results suggest particularly low female labor participation (only 11.3%) in the private sector in Bahrain. However, if we restrict our analysis to citizens only, Bahraini women's labor participation rate — as a percentage of total Bahrainis employed — stands much higher at 30% in the private sector and 48% in the public sector. Between 2005 and 2016, the proportion of employed Bahraini nationals that were women increased in both the private and public sector. Interestingly, although the total proportion of employees in the private sector that were Bahraini nationals declined during this same period, the female proportion of total Bahrainis employed in the private sector increased.

Participation among Bahrainis broken down by economic activity and gender, as shown in Table II, reveals that the finance sector saw the highest participation rate for Bahraini women, followed by the construction sector. Between 2005 and 2016, the proportion of Bahrainis working in finance averaged about 61%. Bahraini women as a percent of total Bahrainis employed in finance increased from a low of 31.84% in the first quarter of 2005 to 35.51% in the first quarter of 2016. The proportion of national women employed in the sector peaked at 38.33% in the fourth quarter of 2010.

Table II: Employed Bahrainis by Economic Activity.

	Construction	Trade	Finance
Men	8,459	20,209	5,810
Women	3,871	6,159	3,166
Total	12,330	26,368	8,976
Percent Women	31%	23%	35%

Source: Information gathered from Government of Bahrain, "Table A06: Estimated Total Employment by Citizenship and Sector (Construction) 2006–2017" (2017); Government of Bahrain, "Table A07: Estimated Total Employment by Citizenship and Sector (Trade) 2006–2017" (2017); and Government of Bahrain, "Table A10: Estimated Total Employment by Citizenship and Sector (Finance) 2006–2017" (2017).

3.1.2 *Kuwait*

Under Decree no. 1104/2008, Kuwait enforced nationalization quotas on specific sectors, including the banking, real estate, education, and manufacturing sectors. The aggregate labor force participation rates in Kuwait depicted in Table III seemingly show lower levels of women's inclusion in the Kuwaiti workforce; a statistic consistent with conventional wisdom and assumptions in the literature. When we lump all employed persons together — nationals and expatriates — women

Table III: Distribution of Employed Population in the Public and Private Sector by Nationality and Gender, Kuwait, 2016.

	All (Nationals and Foreigners)		Nationals	
	Governmental Jobs	*Private*	*Governmental Jobs*	*Private*
Men	249,503	1,335,086	164,940	31,767
Women	195,977	144,133	152,229	26,189
Total	445,480	1,479,219	317,169	57,956
Percent Women	44%	9.74%	48%	45.19%

Source: Government of Kuwait, "Workers in the Private Sector by Economic Activity Registered in the License and Nationality (Kuwaiti/non-Kuwaiti) and Gender", The Public Authority for Civil Information (2016).

constitute only 9.74% of the workforce in the private sector in 2016 and 44% in the public sector. However, examining employed nationals alone suggests a different picture. Kuwaiti women make up 45% of the total national Kuwaiti workforce in the private sector and 48% in the public sector, percentages clearly reflecting greater gender equality in the Kuwaiti workforce.

Table IV presents the distribution of Kuwaitis by gender across six different economic activities. Data broken down by economic activity reflect high female labor force participation among sectors targeted by nationalization. For example, Kuwaiti women make up almost half of all Kuwaitis employed in the manufacturing, construction, and finance sectors. Kuwaiti women also outnumber Kuwaiti men in the education and retail sectors, where they make up 82% and 54% of all Kuwaitis employed in those sectors, respectively.

Table IV: Employed Kuwaitis by Economic Activity (Government and Non-Government Sectors), 2016.

	Kuwaiti Nationals			
Sector	Men	Women	Total	Percent Women
Manufacturing	2,865	2,650	5,515	48%
Construction	11,084	10,005	21,089	47%
Retail and Wholesale	8,983	10,513	19,496	54%
Financial Intermediaries	4,878	3,919	8,797	45%
Real Estate	11,278	5,100	16,378	31%
Education	457	2,094	2,551	82%
Transport and Storage	2,595	1,421	4,016	35%

Source: Government of Kuwait, "Employment by Economic Activity and Sector", The Public Authority for Civil Information (2016).

3.1.3 *Oman*

Omanization was first introduced in 1988 as a national priority and has remained integral to every five-year strategic vision plan issued since. The 2012 Omani Labor Law clearly states that employers must employ Omani citizens to the maximum extent possible.[58] Moreover, the law states that work permits to non-Omanis are granted only if there are no qualified Omanis to fill the vacant posts, and if the employer has met the specified Omanization targets. Omanization targets are set as follows: 60% for jobs in transport, storage, and communications; 45% for

---

[58] Oman Govt, "Labour Law" (2012).

positions in finance, insurance, and real estate; 35% for industrial jobs; 30% for hotels and restaurants; 20% for retail jobs; and 15% for contracting positions.[59] There are also occupations that are limited to Omani citizens only, such as department managers, primary school teachers, nurses, general car mechanics, and technicians.[60] Oman's extensive labor market nationalization efforts have allowed more Omani women to join the workforce in both the public and private sector. In fact, Omani women made up 41% of the total citizen workforce in the public sector in 2015 and 24% of the private sector.[61]

As of December 2016, 15,434 Omanis worked in financial intermediation in the private sector; 6,953 of which were women, constituting 45% of nationals employed in the sector.[62] The financial sector was the source of economic activity with the third-highest proportion of employed Omani women working in the private sector (12.85%), followed closely by real estate at 12.5%. The economic activities with the largest proportions of employed national women in the private sector were wholesale/retail and trade (21%) and construction (19.23%).

The banking industry — one of the more nationalized sectors in the economy — also has a high percentage of Omani females. In 2016, Omani women accounted for 43.8% of all Omanis employed in local banks and 39.1% of all Omanis working in foreign banks in Oman. Similarly, the percentage of Omani women working in the telecommunication sector represents 27.8% of all employed nationals. Both sectors have higher percentages of Omani females than the overall average for women in the private sector. This, in turn, shows how Omanization helped facilitate the entry of more Omani women into the workforce.

3.1.4 *United Arab Emirates*

The UAE's nationalization efforts began targeting the private sector in 2000, while Emiratization of the public sector dates back to the 1990s.[63] In 2012, Sheikh Mohamed bin Rashid Al Maktoum, UAE's prime minister and Dubai's ruler, issued a law requiring government departments and public companies to have female representation on their corporate boards.[64] UAE nationalization policies have targeted "relatively high-skill job classifications", including financial services and aerospace engineering.[65] As in other GCC countries, implementation has been inconsistent, owing in part to social constraints such as jobs deemed socially unacceptable for citizens, or nationals lacking the necessary skills to perform them. However, one sector that has seen relatively consistent implementation is the banking sector, which has also been the focus of more rigorous nationalization efforts. In 2004, the government required all bank manager positions to be

---

[59] Anon., "Omanisation Programme and Policy", *Oman Information Center* (n.d.).

[60] Das and Gokhale, "Omanization Policy and International Migration in Oman", *Middle East Institute*, 2 Feb. 2010.

[61] Oman Govt, "Oman's 2016 Statistical Yearbook", *National Centre for Statistics and Information* 44 (2016).

[62] Oman Govt, "Monthly Statistical Bulletin", *National Centre for Statistics and Information* 28 (Feb. 2017).

[63] Matherly and Hodgson, "Implementing Employment Quotas to Develop Human Resource Capital: A Comparison of Oman and the UAE", *International Journal of Liberal Arts and Social Science* 2.7 (2014), pp. 75–90; Fasano and Goyal, "Emerging Strains in GCC Labor Markets".

[64] Sleiman, "GCC Women Take Charge as Economies Shift", *Al Arabiya English*, 17 Apr. 2014.

[65] Barnett, Malcolm, and Toledo, "Shooting the Goose that Lays the Golden Egg: The Case of UAE Employment Policy", *Journal of Economic Studies* 42.2 (2015), p. 299. According to the authors, "the long-term target appears to be 55%, according to a 2009–2012 Emiratization plan developed by the UAE State Audit Institution" (p. 287). Also see: Matherly and Hodgson, "Implementing Employment Quotas", p. 77.

UAE nationals.[66] In addition, the government issued a requirement that banks increase their employment of citizens by 4% per year, and by 5% per year in the insurance sector.[67]

Table V presents the distribution of employees by gender and nationality in the financial and insurance services sector for the Emirate of Dubai between 2007 and 2014. During this period, nationals averaged 18.62% of the sector's employees and the majority of these were female. Emirati women accounted for a low of 62.44% of Emiratis employed in the sector in 2007, to a high of 69.78% in 2014. Moreover, male non-citizen employees in the Emirate of Dubai working in the financial sector outnumbered female non-citizen employees approximately three to one. However, the reverse was true among national employees: the number of female citizens employed in finance was nearly twice that of male citizens during this period.

According to the 2012 Labor Force Survey, 14.01% of employed citizen females in the Emirate of Dubai worked in the financial intermediation sector, compared to only 4.36% of citizen males.[68] In fact, this sector was the second-largest employer of Emirati women behind public administration and defense, where 41.18% of employed Emirati females worked. We see a similar pattern in Abu Dhabi and Oman. In 2008, 16.83% of those employed in financial intermediation in the Emirate of Abu Dhabi were nationals; of those, 42% were female. In 2011, the proportion of female citizens reached 51.52% of the 16.92% total Emiratis employed in the financial sector. As in Dubai, citizen female employees in 2011 represented double the number of non-citizens in the financial and insurance services sector in Abu Dhabi. In addition, for most years, financial and insurance services accounted for the third-

Table V: Distribution of Employees in Financial and Insurance Services, Emirate of Dubai.

Year	Total Employed	% Emirati of Total	% Female of Total Emirati	Citizens % Male of Total (Frequency)	Citizens % Female of Total (Frequency)
2007	33,979	19.67	62.44	7.39 (2,510)	12.28 (4,173)
2008	35,430	19.89	63.17	7.33 (2,596)	12.57 (4,452)
2009	34,891	19.65	62.56	7.36 (2,567)	12.29 (4,289)
2010	34,354	21.05	66.61	7.03 (2,414)	14.02 (4,816)
2011	35,482	18.50	65.75	6.34 (2,248)	12.16 (4,316)
2012	34,643	16.82	63.97	6.06 (2,100)	10.76 (3,728)
2013	36,445	16.83	68.26	5.34 (1,946)	11.49 (4,186)
2014	36,882	16.54	69.78	5.00 (1,843)	11.54 (4,257)

Source: Government of Dubai, *Financial Services* (Dubai: Dubai Statistics Center, 2007–2015).

---

[66] Ibid.

[67] Barnett, Malcolm, and Toledo, "Shooting the Goose", p. 287.

[68] Govt of Dubai, "Percentage Distribution of Employed 15 Years and Over by Nationality, Sex and Economic Activity – Emirate of Dubai" (2012).

Table VI: Percentage Distribution of Employed Nationals in Financial and Insurance Services Sector, Emirate of Abu Dhabi.

Year	Total Percent Employed (All)	Citizens		
		% Total	% Male	% Female
2008	1.4	2.5	1.8	5.0
2011	1.4	2.8	1.8	5.9
2012	2.0	2.8	1.5	6.5
2013	1.6	2.7	1.6	6.1
2014	2.0	3.2	1.1	8.8
2015	2.3	3.8	2.1	8.0

Source: Government of Abu Dhabi, *Statistical Yearbook of Abu Dhabi* (Abu Dhabi: Statistic Centre – Abu Dhabi, 2011–16).

highest share of employed Emirati women, behind public administration, defense, and education.[69]

Table VI shows the percentage distribution for employed citizens working in the financial sector for the Emirate of Abu Dhabi. The table reports the proportion of the total work force, including nationals and non-nationals, employed in financial and insurance services as well as the proportion of nationals by gender. In 2008, 2.5% of employed nationals in the Emirate of Abu Dhabi worked in this sector. When we look at the distribution by gender, we can see that 5% of employed citizen women worked in this sector, compared to 1.8% of Emirati men. The share of employed nationals working in this sector increased — though not monotonically — from 2.5% in 2008 to 3.8% in 2015. Much of this increase was due to a growing proportion of employed citizen women working in the financial services sector, the share increasing from 5% in 2008 to 8% of all employed Emirati women in 2015, compared to only 2.1% of employed national men. In fact, the proportion of employed national women in the financial services sector was three to four times higher than the proportion of employed male citizens.

Finally, Table VII presents the distribution of employed nationals by gender and sector in Abu Dhabi. For the three years for which data were available, women were employed in the private sector and in joint private-public entities at higher rates than their male counterparts.

Table VII: Distribution of National Employees by Sector, Emirate of Abu Dhabi.

		All Citizens	Male	Female
2012	Government	86.4%	87.6%	83.1%
	Private Sector	5.7	5.4	6.6
	Joint (Government & Private)	5.4	5.0	6.4
2013	Government	88.1%	89.2%	84.8%
	Private Sector	6.0	5.3	8.2
	Joint (Government & Private)	5.3	4.9	6.5
2014	Government	81.9%	82.6%	80.0%
	Private Sector	4.9	4.8	5.2
	Joint (Government & Private)	12.0	11.9	12.4

Source: Government of Abu Dhabi, *Statistical Yearbook of Abu Dhabi* (Abu Dhabi: Statistic Centre – Abu Dhabi, 2013–15).

---

[69] It is interesting to note that while analyzing the raw data, we found that defense actually has the highest proportion of employed national women in the public sector. One third of the Ministry of Defense's employees are women [Al-Kuttab, "Defending the Homeland is no more a Male-Dominated Sector in the UAE", *Khaleej Times*, 22 Dec. 2017]. While we do not have an explanation for why so many women work in defense, it is certainly worth investigating in future research.

3.1.5 *Qatar*

From 1997, nationalization efforts in Qatar aimed to increase the proportion of citizens in all sectors to 20%.[70] According to Hertog, however, "Qatarization policies have been pursued even less systematically than equivalent policies in the UAE and Kuwait".[71] However, in 2015, the Minister of Administrative Development stated that Qatar aimed to increase the number of nationals in the public sector such that by 2026 nine of ten jobs will be held by Qataris.[72] Interestingly, the number of Qataris employed in the private sector has actually declined, with only around one or two percent of citizens working in the private sector.[73]

According to statistics from the Ministry of Development Planning and Statistics, 25% of employees in the financial and insurance activities sector were Qatari in 2013, of which 54% were female. Moreover, although few Qataris work in the private sector, women made up around 40% of the Qataris employed in 2013, but around 35% in 2014 and 2015. In 2008, Qatari nationals comprised about 0.50% of private sector employees, of which 26.34% were female. The private sector accounted for about 5% of total female national employment in 2008, but by 2014, the private sector accounted for 14.41% of all employed Qatari females compared to 10.66% for Qatari males. However, the private sector's share declined again to 12.10% of employed female citizens in 2015.

3.1.6 *Saudi Arabia*

In contrast to the previous cases, the effects of nationalization efforts in Saudi Arabia are difficult to examine due to restrictions on women's mobility and a largely native workforce. Thus, in many ways, Saudi Arabia is an outlier in terms of the argument we introduce here. Saudization was introduced in the 1990s to increase national employment and reduce reliance on foreign workers. During the first phase of Saudization, identical national employment quotas were set for all employers, regardless of the size or activity of their firms.

The second phase of Saudization brought *Nitaqat*, which set quotas for citizen employment relative to the economic activity and size of the corporation.[74] Each company was categorized as platinum, green, yellow, or red, in accordance to its compliance towards Saudization quotas, and then was granted special hiring privileges based on its color-tier.[75] *Nitaqat* was completed in 2013 with the Saudi government now planning a third phase, which explicitly includes a provision for increasing female employment to 28% by 2020.[76]

Data on Saudi employment from the years 1999 to 2016, presented in Table VIII, suggest that nationalization policies helped incorporate larger numbers of Saudi women into the workforce. Although Saudi women's share of total national employment remains small at approximately 16%,

---

[70] Felder and Vuollo, "Qatari Women in the Workforce", p. 5.

[71] Hertog, "Arab Gulf States", p. 22.

[72] Khatari, "Minister: Qataris to Comprise 90 Percent of Public Sector by 2026", *Doha News*, 27 Dec. 2015.

[73] Felder and Vuollo, "Qatari Women in the Workforce", pp. 27–8, estimate 2%. The Ministry of Development Planning and Statistics reports less than one percent of Qatari nationals working in the private sector with 0.95%, 0.85%, and 0.77% in 2013, 2014, and 2015, respectively [Govt of Qatar, *Bulletin: Labor Force Statistics, 2015* (2015)].

[74] According to Randeree "the requirements ranged from 6% in the construction sector, to 19% in media, and 49% in the banking sector" [Randeree, "Workforce Nationalization", p. 13].

[75] Anon., "Mobility: Immigration Alert", *Ernst and Young* (2016).

[76] Anon., "Saudi Arabia Wants More Women in the Workforce by 2020: Ministry of Labor", *Albawaba*, 22 Mar. 2017.

Table VIII: Female and Male National Employment in Saudi Arabia, 1999–2016.

Year	Employed Nationals			Percent Women
	Males	Females	Total	
1999	2,247,720	347,370	2,595,090	13.39%
2000	2,352,092	351,279	2,703,371	12.99%
2001	2,413,780	363,193	2,776,973	13.08%
2002	2,480,225	364,366	2,844,591	12.81%
2003	2,594,493	396,844	2,991,337	13.27%
2004	2,708,760	429,323	3,138,083	13.68%
2005	2,823,028	461,801	3,284,829	14.06%
2006	2,937,295	494,279	3,431,574	14.40%
2007	3,082,301	502,456	3,584,757	14.02%
2008	3,185,417	493,183	3,678,600	13.41%
2009	3,332,628	505,340	3,837,968	13.17%
2010	3,411,801	543,406	3,955,207	13.74%
2011	3,538,669	604,402	4,143,071	14.59%
2012	3,750,781	646,590	4,397,371	14.70%
2013	3,989,632	727,495	4,717,127	15.42%
2014	4,120,467	805,717	4,926,184	16.36%
2015	4,159,744	816,361	4,976,105	16.41%
2016	4,185,853	835,726	5,021,579	16.64%

Source: Government of Saudi Arabia, "Labor Force" (2017).

the percentage change in the number of Saudi women employed across this period is significant. In 1999, only 347,370 Saudi women were registered as employed compared to 835,726 in 2016, reflecting a dramatic increase of 140%. During the second phase of *Nitaqat*, the number of employed Saudi women increased from 543,406 in 2010 to 805,717 in 2014, representing a 48% growth in labor force participation. The percentage change in the number of Saudi men employed from 1999 to 2016 was 86%, while the percentage change during *Nitaqat's* second phase was 21% — significantly lower than the percentage change in the employment of Saudi women.

Although the number of employed women drastically increased with the introduction of nationalization policies, the number of unemployed Saudi women also increased, indicating that as more national women joined the labor force, a significant proportion of women were unable to find jobs. Unemployment rates among Saudi women in 2016 stood at 34%, up from 16% in 1999, while unemployment rates among Saudi men fell from 6.8% in 1999 to 5.7% in 2016.[77] This in turn indicates that although oil-driven nationalization policies may have helped in incorporating more Saudi women into the workforce, this positive impact was less than the other GCC countries. Consequently, Saudi Arabia, despite being its mainstream invocation as emblematic of the GCC, is, in fact, not a good proxy for examining regional patterns.

4 Conclusion

When discussing the gendered nature of the resource curse, Ross succinctly states that "the benefits of oil booms usually go to men".[78] In this paper, we argue that the oil-rich GCC countries appear to contradict this conclusion. More specifically, we contend that labor market

---

[77] Govt of Saudi Arabia, "Labor Force" (2017).
[78] Ross, *The Oil Curse*, p. 111.

nationalization efforts coincided with a period in which female nationals were making substantial gains in educational attainment, especially relative to their male counterparts. This confluence of factors created an opportunity that led to an increased number of qualified female citizens joining the labor market. Both women's gains in education and nationalization policies are directly linked to the Gulf states' oil wealth and oil-led modernization strategy. In this way, oil was a blessing — not a curse — for women's economic participation.

Although the lack of regular and standardized time-series data by sector and economic activity constrains our ability to conduct a more systematic examination,[79] we believe that what data are available represent a key challenge to the literature connecting oil rents to gender inequality. However, it is unlikely that our conclusions would extend to other oil-rich states given the complex nature of rentierism in the Gulf states, including the demographic and cultural specificities of the GCC. Furthermore, it may be too early to appreciate the full impact of nationalization policies on female labor force participation as they have only been implemented in earnest relatively recently. Nevertheless, these policies appear to have created significant opportunities for women to enter the workforce and, importantly, their employment trends seems to coincide with sectors targeted by such policies.

Labor nationalization efforts are just one policy tool that GCC countries can and do use to facilitate entry into the workforce. While labor nationalization explains the overarching incentive for citizen women to join the workforce, other policies have encouraged them as well. Many such policies directly relate to motherhood and balancing work and family life, including mandating paid maternity leave. In addition, many of the recent vision statements recognize that female citizens are an untapped resource and include explicit reference to women and their importance in the workforce. The Qatar National Development Strategy, for example, emphasizes that "women's empowerment is both a goal and an enabler to achieve the aspirations of *Qatar National Vision 2030*".[80]

Despite these efforts, however, a range of obstacles to women entering and remaining in the labor force persists. Traditional norms and attitudes, in particular, continue to stand out as a major challenge for women in the labor force.[81] Many women also decide not to return to work after having children. Conservative Arab society still emphasizes women's primary commitment to home and family, and most of these states promote both women's participation in the workforce and large families in a bid to increase the native population. Certainly, these ambitions can cause tensions for working women with children. Nevertheless, even over a short period, attitudes of women —and towards women — appear to be quickly evolving. Younger female citizens show greater willingness to enter occupations in a much broader range of sectors, including engineering and other professional occupations, than older generations of GCC women.[82]

While GCC states have tasked themselves with integrating women in the workforce in the millennium, not all of them have dedicated equal resources or political will to this endeavor. For instance, there are different attitude toward the hours women should work across individual states. While Kuwait, Bahrain, and the UAE do not seem to restrict the hours women can work, and Kuwait even offers extra security measures and special transportation for women working

---

[79] On this issue, see: Barnett; Malcolm; and Toledo, "Shooting the Goose", p. 290; Rutledge et al., "Women, Labour Market Nationalization Policies", p. 189–90.

[80] Govt of Qatar, *Qatar National Development Strategy 2011–2016: Towards Qatar National Vision 2030* (2011), p. 178.

[81] Willen et al., "Power Women in Arabia"; Mitchell et al., "In *Majaalis al-Hareem*: The Complex Professional and Personal Choices of Qatari Women", *DIFI Family Research and Proceedings* (2015); Rutledge et al., "Women, Labour Market Nationalization Policies".

[82] Felder and Vuollo, "Qatari Women in the Workforce".

from 10:00pm to 7:00am, Oman restricts women from working between the hours of 9:00pm to 6:00am. These restrictions have led to questions about the genuine commitment of certain states to ensuring women's access to the public realm.

Finally, although GCC women have made substantial gains in terms of labor force participation over the last two decades, women are not equally represented at higher levels of civil society or in business.[83] According to an A.T. Kearney Report, women make up only 9% of senior positions in Fortune 500 companies in the GCC, compared to 35%, 29%, 26%, and 21% in the Americas, Africa, Asia, and Europe, respectively.[84] Moreover, the data we present here cannot speak to whether women are actually advancing in the workplace, becoming managers and executives in the various sectors in which they increasingly dominate. Determining and dismantling the obstacles to advancement and promotion will be crucial for truly improving women's economic empowerment and social inclusion across the GCC.

This also raises the question of whether the path to gender empowerment is inexorably paved through women's economic opportunities. Ultimately, women's entry into the workforce and their levels of labor force participation are an important, but also a limited, tool for improving their representation in other arenas, including the political sphere.[85] Beyond just increasing the proportion of women in the labor force, achieving gender parity requires multifaceted efforts to tackle barriers across society and "propel women to reach the leadership positions they seek and deserve".[86] There is clearly a need to formulate policies to address cultural and structural obstacles to female employment in various sectors that will, in turn, help "normalize" female employment further in the public, and especially in the private, sector.[87] Thus, although national women living in the GCC seem to be doing very well on certain economic indicators, this is not the case for other dimensions of gender equality such as those capturing their selfhood and autonomy in the political and social realms. Furthermore, with oil prices falling and a history of women losing out first in public spending cuts, it remains to be seen whether their impressive gains in economic and political spaces in the GCC will last.

Bibliography

1 Primary sources

Anon., *The Global Gender Gap Report* (Geneva: World Economic Forum, 2015), available online at www3. weforum.org/docs/GGGR2015/cover.pdf.

———, "Mobility: Immigration Alert", *Ernst and Young* (2016), available online at www.ey.com/ Publication/vwLUAssets/Saudi_Arabia_Introduces_new_Labor_Market_Test_and_increased_visa_ application_fees_impacting_immigration_timing_and_costs/$FILE/Saudi%20Arabia%20-% 20Introduction%20of%20new%20Labor%20Market%20Test%20and%20increased%20visa% 20application%20fees%20impacting%20immigration%20timing%20and%20costs.pdf.

———, "Omanisation Programme and Policy", *Oman Information Center* (n.d.), available online at www. omaninfo.com/manpower-and-employment/omanisation-programme-and-policy.asp.

---

[83] Ibid., p. 15.

[84] Willen et al., "Power Women in Arabia", p. 4.

[85] See Iversen and Rosenbluth, "Work and Power: The Connection between Female Labor Force Participation and Female Political Representation", *The Annual Review of Political Science* 11 (2008), pp. 479–95, for an excellent review of the literature examining the relationship between economic and political empowerment.

[86] Willen et al., "Power Women in Arabia", p. 8.

[87] Rutledge et al., "Women, Labour Market Nationalization Policies".

Government of Abu Dhabi, *Statistical Yearbook of Abu Dhabi* (Abu Dhabi: Statistic Centre – Abu Dhabi, 2018), available online at www.scad.ae/en/pages/GeneralPublications.aspx?pubid = 79&themeid = 7.

———, "Demographic & Social Indicators", Statistics Centre – Abu Dhabi (2008), available online at www. scad.ae/Release%20Documents/Demographic%20and%20Social%20Indicators%20-%20Labour% 20Force.pdf.

Government of Bahrain, "Table A1: Estimated Total Employment (Male) by Citizenship and Sector: 2006– 2017" (2017), available online at http://blmi.lmra.bh/2017/12/data/lmr/Table_A1.pdf.

———, "Table A06: Estimated Total Employment by Citizenship and Sector (Construction) 2006–2017" (2017), available online at http://blmi.lmra.bh/2017/12/data/lmr/Table_A06.pdf.

———, "Table A07: Estimated Total Employment by Citizenship and Sector (Trade) 2006–2017" (2017), available online at http://blmi.lmra.bh/2017/12/data/lmr/Table_A07.pdf.

———, "Table A10: Estimated Total Employment by Citizenship and Sector (Finance) 2006–2017" (2017), available online at http://blmi.lmra.bh/2017/12/data/lmr/Table_A10.pdf.

Government of Dubai, "Percentage Distribution of Employed 15 Years and Over by Nationality, Sex and Economic Activity – Emirate of Dubai" (2012), www.dsc.gov.ae/Report/DSC_LFS_2012_02_05.pdf.

———, *Financial Services* (Dubai: Dubai Statistics Center, 2007–15), available online at www.dsc.gov.ae/ en-us/Themes/Pages/Financial-Services.aspx?Theme = 32.

Government of Kuwait, "Employment by Economic Activity and Sector", The Public Authority for Civil Information (2016), available online at www.paci.gov.kw/stat/default.aspx.

———, "Workers in the Private Sector by Economic Activity Registered in the License and Nationality (Kuwaiti/non-Kuwaiti) and Gender", The Public Authority for Civil Information (2016), available online at http://stat.paci.gov.kw.

Government of Oman, "Labour Law" (2012), available online at www.oman.om/wps/wcm/connect/ ac78dc4f-69f0-4ddd-ad36-eefb357a43f6/Omani + Labour + Law.pdf?MOD = AJPERES.

———, *Oman's 2016 Statistical Yearbook* (Muscat: National Centre for Statistics and Information, 2016), available online at www.ncsi.gov.om/Elibrary/LibraryContentDoc/bar_Statistical%20Year%20Book% 202016%20_6f43e0e6-592a-43a1-b1b6-82c32c046aa9.pdf.

———, "Monthly Statistical Bulletin", *National Centre for Statistics and Information* 28 (2017), available online at www.ncsi.gov.om/Elibrary/LibraryContentDoc/bar_MSBFeb20173_e586cf2d-0f85-4d12- b275-3eb9de5f2743.pdf.

Government of Qatar, *Bulletin: Labor Force Statistics, 2015* (Doha: Ministry of Development Planning and Statistics, 2015), available online at www.mdps.gov.qa/en/statistics/Statistical%20Releases/Social/ LaborForce/2015/LaborForce2015.pdf.

———, *Qatar National Development Strategy 2011–2016: Towards Qatar National Vision 2030* (Doha: Qatar General Secretariat for Development Planning, 2011), available online at www.mdps.gov.qa/en/ knowledge/HomePagePublications/Qatar_NDS_reprint_complete_lowres_16May.pdf.

Government of Saudi Arabia, "Labor Force" (2017), available online at www.stats.gov.sa/en/814.

Gulf Research Center, "Demographic and Economic Module: Data on Population Stocks", Gulf Labour Markets, Migration, and Population Programme (2018), available online at http://gulfmigration.eu/ glmm-database/demographic-and-economic-module.

World Bank, "World Development Indicators 1990–2014" (2016), available online at http://databank. worldbank.org/data/home.aspx.

———, "Gender and Development in the Middle East and North Africa: Women in the Public Sphere" (2004), available online at http://documents.worldbank.org/curated/en/976361468756608654/pdf/ 281150PAPER0Gender010Development0in0MNA.pdf.

2 Secondary sources

Al Kuttab, Jasmine, "Defending Homeland Is No More a 'Male-Dominated' Sector in UAE", *Khaleej Times*, 22 December 2017, available online at www.khaleejtimes.com/nation/abu-dhabi/defending-homeland- is-no-more-a-male-dominated-sector-in-uae.

Al-Majdali, Fauzi, "National Labor Quotas Increased in Private Sector", *Kuwait Times*, 8 August 2014, available online at http://news.kuwaittimes.net/national-labor-quotas-increased-private-sector.

Anon., "Gulf Arab States May See Unemployment Rise, Labour Reforms Needed-IMF", *Reuters*, 20 November 2013, available online at www.reuters.com/article/imf-gulf-jobs/gulf-arab-states-may-see- unemployment-rise-labour-reforms-needed-imf-idUSL5N0J51IQ20131120.

————, "Nine Top Officials Sacked after Dubai Ruler's Spot Check Reveals Empty Desks", *The Telegraph*, 30 August 2016, available online at www.telegraph.co.uk/news/2016/08/30/dubais-ruler-orders-management-shake-up-after-unannouced-inspect.

————, "'Sack All Foreigners by 2020', Says Saudi Government", *The New Arab*, 10 May 2017, available online at www.alaraby.co.uk/english/news/2017/5/10/sack-all-foreigners-by-2020-says-saudi-government.

————, "Saudi Arabia Wants More Women in the Workforce by 2020: Ministry of Labor", *Albawaba*, 22 March 2017, available online at www.albawaba.com/business/saudi-arabia-women-nitaqat-953282.

Barnett, Andy H.; Michael Malcolm; and Hugo Toledo, "Shooting the Goose that Lays the Golden Egg: The Case of UAE Employment Policy", *Journal of Economic Studies* 42.2 (2015), pp. 285–302.

Buttorff, Gail; Bozena Welborne; and Nawra Al Lawati, "Measuring Female Labor Force Participation in the GCC", *Baker Institute Issue Brief*, 18 January 2018, available online at www.bakerinstitute.org/media/files/files/1fe4b913/bi-brief-011818-wrme-femalelabor.pdf.

Das, Kailash C. and Nilambari Gokhale, "Omanization Policy and International Migration in Oman", *Middle East Institute*, 2 February 2010, available online at www.mei.edu/content/omanization-policy-and-international-migration-oman.

DeBoer, Jennifer and Ashley Ater Kranov, "Key Factors in the Tertiary Educational Trajectories of Women in Engineering: Trends and Opportunities in Saudi Arabia, the GCC, and Comparative National Settings", *Science and Technology in the Gulf States*, edited by Afreen Siddiqi and Laura Diaz Anadon (Berlin: Gerlach Press, 2017), pp. 56–88.

Fasano, Ugo and Rishi Goyal, "Emerging Strains in GCC Labor Markets", International Monetary Fund, working paper 04.71 (2004), available online at www.imf.org/en/Publications/WP/Issues/2016/12/30/Emerging-Strains-in-GCC-Labor-Markets-17286.

Felder, Dell and Mirka Vuollo, "Qatari Women in the Workforce", *Rand-Qatar Policy Institute*, working paper no. WR-612-Qatar (2008), available online at www.rand.org/pubs/working_papers/WR612.html.

Forstenlechner, Ingo and Emilie Rutledge, "Unemployment in the Gulf: Time to Update the 'Social Contract'", *Middle East Policy* 17.2 (2010), pp. 38–51.

————, "The GCC's 'Demographic Imbalance': Perceptions, Realities and Policy Options", *Middle East Policy* 18.4 (2011), pp. 25–43.

Herb, Michael, *The Wages of Oil: Parliaments and Economic Development in Kuwait and the UAE* (Ithaca: Cornell University Press, 2014).

Hertog, Steffen, "A Comparative Assessment of Labor Market Nationalization Policies in the GCC", *National Employment, Migration and Education in the GCC*, edited by Steffen Hertog (Berlin: Gerlach Press, 2012), pp. 65–106.

————, "Arab Gulf States: An Assessment of Nationalisation Policies", *Gulf Labour Markets and Migration*, research paper no. 1/2014 (2014), available online at http://cadmus.eui.eu/handle/1814/32156.

Iverson, Torben and Frances Rosenbluth, "Work and Power: The Connection between Female Labor Force Participation and Female Political Representation", *The Annual Review of Political Science* 11 (2008), pp. 479–95.

Jones Luong, Pauline and Erika Weinthal, *Oil is not a Curse: Ownership Structure and Institutions in Soviet Successor States* (Cambridge: Cambridge University Press, 2010).

Kerr, Simeon, "Saudi Arabia Cuts Public Sector Bonuses in Oil Slump Fallout", *The Financial Times*, 27 September 2016, available online at www.ft.com/content/765898e0-8482-11e6-8897-2359a58ac7a5.

Khatri, Shabina S., "Minister: Qataris to Comprise 90 Percent of Public Sector by 2026", *Doha News*, 27 December 2015, https://dohanews.co/minister-qataris-comprise-90-percent-public-sector-2026/.

Liou, Yu-Ming and Paul Musgrave, "Oil, Autocratic Survival, and the Gendered Resource Curse: When Inefficient Policy Is Politically Expedient", *International Studies Quarterly* 60.3 (2016), pp. 440–56.

Matherly, Laura L. and Sasha Hodgson, "Implementing Employment Quotas to Develop Human Resource Capital: A Comparison of Oman and the UAE", *International Journal of Liberal Arts and Social Science* 2.7 (2014), pp. 75–90.

Mitchell, Jocelyn Sage; Christina Paschyn; Sadia Mir, Kirsten Pike; and Tanya Kane, "In *Majaalis al-Hareem*: The Complex Professional and Personal Choices of Qatari Women", *DIFI Family Research and Proceedings* 2015–4, available online at http://dx.doi.org/10.5339/difi.2015.4.

Randeree, Kasim, "Workforce Nationalization in the Gulf Cooperation Council States", *CIRS Occasional Paper* 9 (Doha: Center for International and Regional Studies, Georgetown University in Qatar, 2012).

Ridge, Natasha, *Education and the Reverse Gender Divide in the Gulf States: Embracing the Global, Ignoring the Local* (New York: Teachers College Press, 2014).

Ross, Michael L., "Oil, Islam, and Women", *American Political Science Review* 102.1 (2008), pp. 107–23.

———, *The Oil Curse: How Petroleum Wealth Shapes the Development of Nations* (Princeton: Princeton University Press, 2012).

Ruppert, Elizabeth "Managing Foreign Labor in Singapore and Malaysia: Are there Lessons for GCC Countries?", *World Bank Publications* no. 2053 (1999), available online at https://elibrary.worldbank.org/doi/abs/10.1596/1813-9450-2053.

Rutledge, Emilie; Fatima Al Shamsi; Yahia Bassioni; and Hend Al Sheikh, "Women, Labour Market Nationalization Policies and Human Resource Development in the Arab Gulf States", *Human Resource Development* 14.2 (2011), pp. 183–98.

Rutledge, Emilie and Fatima Al Shamsi, "The Impact of Labor Nationalization Policies on Female Participation Rates in the Arab Gulf", *Women, Work and Welfare in the Middle East: The Role of Socio-demographics, Entrepreneurship and Public Policies*, edited by Nadereh Chamlou and Massoud Karshenas (London: Imperial College Press, 2016), pp. 525–51.

Schake, Kori, "The Myth of the Resource Curse", *Defining Ideas*, a Hoover Institute Journal, 25 October 2012, available online at www.hoover.org/research/myth-resource-curse.

Shalaby, Marwa, "The Paradox of Female Economic Participation in the Middle East and North Africa", *Baker Institute Issue Brief*, 3 July 2014, available online at www.bakerinstitute.org/research/female-economic-participation-middle-east.

Sleiman, Mirna, "GCC Women Take Charge as Economies Shift", *Al Arabiya English*, 17 April 2014, available online at http://english.alarabiya.net/en/perspective/features/2014/04/17/Women-edge-into-Gulf-boardrooms-as-economies-societies-shift.html.

Stasz, Cathleen; Eric R. Eide; and Francisco Martorell, *Post-Secondary Education in Qatar: Employer Demand, Student Choice, and Options for Policy* (Santa Monica, CA: RAND Corporation, 2007), available online at www.rand.org/pubs/monographs/MG644.html.

Swan, Melanie, "Emiratisation Brings Greater Equality for UAE Women", *The National*, 29 April 2013, available online at www.thenational.ae/news/uae-news/emiratisation-brings-greater-equality-for-uae-women.

Tsujigami, Namie, "Higher Education and Changing Aspirations of Women in Saudi Arabia", *Higher Education Investment in the Arab States of the Gulf: Strategies for Excellence and Diversity*, edited by Dale Eickelman and Rogaia Abu Sharaf (Berlin: Gerlach Press, 2017), pp. 42–54.

Willen, Bob; Ada Perniceni; Rudolph Lohmeyer; and Isabel Neiva, "Power Women in Arabia: Shaping the Path for Regional Gender Equality", A.T. Kearney, 2016, available online at www.atkearney.com/ … /febfbf5e-7506-4ee9-bea6-5f7fc9cb7f29.

Willoughby, John, "A Quiet Revolution in the Making? The Replacement of Expatriate Labor through the Feminization of the Labor Force in the GCC Countries", working paper no. 2004-18 (Department of Economics, American University, Washington, D.C., 2004), available online at www.researchgate.net/publication/277221553_A_Quiet_Revolution_in_the_Making_The_Replacement_of_Expatriate_Labor_through_the_Feminization_of_the_Labor_Force_in_GCC_Countries.

Zerovec, Mojca and Marike Bontenbal, "Labor Nationalization Policies in Oman: Implications for Omani and Migrant Women Workers", *Asian and Pacific Migration Journal* 20.3–4 (2011), pp. 365–87.

6 The Impact of Oil Rents on Military Spending in the GCC Region

Does Corruption Matter?

Mohammad Reza Farzanegan ⓘD

Abstract: This study shows how the level of corruption matters in the way oil rents affect a state's military spending. Using panel data covering the 1984–2014 period for the Gulf Cooperation Countries (GCC), we find that the effect of oil rents on military budgets depends on the extent of political corruption. Oil rents are negatively associated with military spending of the GCC countries. However this, in turn, is moderated by higher levels of corruption. For comparison, we examine this association in non-GCC countries in the MENA region, finding a positive effect of higher oil rents on military spending: this effect is larger in corrupt polities within non-GCC countries. The intermediary role of corruption in the military-oil nexus is robust, controlling for a set of variables that may affect military spending.

1 Introduction

The Gulf Cooperation Council (GCC) and other countries in the Middle East and North Africa (MENA) have the highest levels of military spending as a percentage of their economies worldwide. According to the World Bank,[1] the average military spending — as a percentage of GDP from 2000 to 2015 in the GCC and non-GCC countries of the MENA region — was 5.7% and 3.5%, respectively. Military spending can have high opportunity costs by lowering the state's ability to spend on provision of public goods, such as education and health. Depending on the forward and backward linkages of the military sector with the rest of the economy, higher spending on the military may have a negative or positive effect on overall economic growth. Therefore, it is important to examine the role of key factors in the allocation of military spending within countries.

How can we explain the relatively high record of spending on the military by GCC/MENA countries? Is it due to the so-called "curse of oil", in which higher oil wealth hinders the region's long-term economic growth through factors such as military spending? Indeed, the GCC and the rest of the MENA region's oil rents, as a percentage of GDP, is the highest for

---

Author's note: I wish to thank Zahra Babar, Mehran Kamrava, Suzi Mirgani, two anonymous referees, and all participants in the Resource Curse in the Gulf working groups at the Center for International and Regional Studies (CIRS) at Georgetown University in Qatar for their helpful comments. I also appreciate the support of Sven Fischer and Jackie Starbird in proofreading the paper.

[1] World Bank, "World Development Indicators" (2017).

Table I: Military spending, oil and corruption: a comparative view

Region	Military expenditure (% of GDP)	Oil rents (% of GDP)	CPIA transparency, accountability, and corruption in the public sector rating (1 = low to 6 = high)
Gulf Cooperation Council (GCC)	5.70	28.88	N/A
MENA, non-GCC	3.50	15.58	2.50
Europe & Central Asia	1.81	1.18	2.72
East Asia & Pacific	1.59	0.78	3.02
Sub-Saharan Africa	1.40	9.92	2.74
Latin America & Caribbean	1.30	3.30	3.44

Note: The three main dimensions assessed for Country Policy and Institutional Assessment (CPIA) index are the accountability of the executive to oversight institutions and of public employees for their performance, access of civil society to information on public affairs, and state capture by narrow vested interests.
Source: World Bank Database "Country Policy and Institutional Assessment".

the average of 2000–15 worldwide (28% and 15%, respectively), followed by Sub-Saharan Africa (9.9%), and Latin America (3.3%).

Do higher oil rents automatically lead to more militarization of states' economies? Are such high records of military spending due to specific domestic socioeconomic, demographic, or institutional factors? To what extent can the internal and external conflict risks and regional military competition explain such a significant allocation of budgets to the military?

We contribute to the literature on the political economy of military spending by taking into account the joint effect of oil rents and political corruption in explaining the military spending of the GCC economies. There are studies that have investigated the independent effects of oil rents and corruption on conflict and military spending. However, to the best of our knowledge, the combined effect of oil rents and corruption on military spending is a neglected area. We fill this gap in our analysis. Table I shows a comparative picture of military spending, oil rents, and levels of transparency, accountability, and corruption of governments in different parts of the world. There is a positive association between higher military spending and level of corruption.

Our hypothesis is that the final effect of oil rents on military spending depends on corruption at the government level.[2] In other words, the effects of oil rents on military spending can change its magnitude and/or signal different levels of political corruption. To test this hypothesis, we use past developments in oil rents and corruption on current developments in military expenditures. We use panel data for country and year fixed-effect regressions for five GCC countries from 1984 to 2014. For comparison, we also examine our hypothesis in the sample of ten non-GCC countries in the MENA region. By controlling for the main drivers of military spending, our results show that the final effects of oil rents on military spending in both samples of countries depend on the level of political corruption. This result is robust for different sets of control variables.

Our estimations show that higher levels of oil rents in the past few years are associated with lower military spending within the GCC countries. This negative association is weaker in those GCC countries that have experienced higher levels of political corruption in the past. This finding

---

[2] We use the International Country Risk Guide (ICRG) corruption index. It is originally from 0 to 6 (most corrupt to least corrupt), but we re-scaled it from 1 to 7 (least corrupt to most corrupt). In the MENA sample, this re-scaled corruption index varies from the minimum of 2 (e.g., Israel in some years) to the maximum of 6 (e.g., Iraq, Lebanon, and Libya in some years).

is in line with a study by Bjorvatn and Farzanegan,[3] showing that higher resource rents may lead to political stability in countries with a relatively strong incumbent (low factional politics). When oil rents increase political stability in such polities, then there will be less pressure for higher spending on military and security, unless higher levels of government corruption encourages such spending. The majority of GCC countries have a profile close to the theoretical model of Bjorvatn and Farzanegan's study. The results of the non-GCC sample of the MENA region follow a similar pattern in the increasing "joint effect" of oil rents and corruption in the past on current military spending. A difference with the GCC estimation is a statistically insignificant positive direct association of oil rents and military spending, which is then amplified by higher corruption in the non-GCC sample. Following the theoretical arguments of Bjorvatn and Farzanegan, higher oil rents may lead to political instability and conflict in countries with a weak incumbent or polities with a high degree of fractionalization. Factional politics is closer to the profile of the non-GCC sample in the MENA region. When oil rents fuel the risk of conflict, there will be higher demands for spending on military and security.[4] Such necessity is amplified by higher political corruption.

There is a possibility that a higher level of military spending (our dependent variable) due to the size of transactions and lower transparency of such spending for national security justifications — especially within the MENA/GCC region — increases the risk of corruption and bribery opportunities (one of our key independent variables). There is lower risk of such reverse feedback from military spending on oil rents (another key independent variable). To reduce the risk of simultaneity bias (as one of the endogeneity reasons), we use a four-year lag of oil rents and corruption (and all other explanatory variables) to explain the within-country changes of military spending.[5] It is possible that a one-year lag would not be sufficient to reduce the risk of simultaneity bias. If decision-makers in government predict the following year's military spending and military projects, they may show less willingness to deal with political corruption, in the hope of benefiting more from forthcoming profits/rents in military contracts and arms imports. Such predication of the following year's military budget by politicians then may undermine our identification strategy. Therefore, we use a higher number of lags with respect to our key explanatory variables.

Our results are also a contribution to the resource curse literature. The literature has discussed several channels for transmission of the curse, including higher risk of conflict and instability and thus overspending on the military. In addition, the oil-rich economies, on average, have weaker-quality democratic institutions, partly due to lack of dependence on tax and social contributions by citizens (fiscal channel). In such institutional contexts, the ruling state allocates more rents to military and security elites for gaining their support in a time of crisis, rather than allocating more in the budget for public education and health. In addition, higher risks of conflict in oil-rich economies (especially in those with factional politics) may push them to allocate more for the military sector to maintain political power.

Some studies show how a higher level of corruption facilitates the militarization of the economy. Our work combines these findings in the literature, and argues that higher levels of

---

[3] Bjorvatn and Farzanegan, "Resource Rents, Balance of Power, and Political Stability", *Journal of Peace Research* 52.6 (2015), pp. 758–73; Bjorvatn and Farzanegan, "Natural-Resource Rents and Political Stability in the Middle East and North Africa", *CESifo DICE Report* 13.3 (2015), pp. 33–7.

[4] For a new empirical evidence on the increasing effect of oil rents on risk of conflict in a worldwide sample see: Farzanegan, Lessmann, and Markwardt, "Natural Resource Rents and Internal Conflicts: Can Decentralization Lift the Curse?", *Economic Systems* 42.2 (2018).

[5] Our results are robust; up to four years of lag for our key independent variables, i.e., oil, corruption, and their interaction term. For control variables, we use a one-year lag in all specifications.

oil rents do not necessarily lead to the enlargement of military budgets. The level of corruption is an important institutional factor that can influence this nexus and shed more light on different experiences of military budgeting within the oil-rich economies. Of course, we have also controlled for other factors, which are correlated with oil rents and corruption on one side, and military spending on the other. Not controlling for such confounding variables may lead to omitted variable bias, which is again another source of the endogeneity issue. We also control for country and year-specific characteristics — such as geography, religion, ethnic and religious fractionalization, cultural attitudes, and historical heritage — which also may be relevant for explaining the size of the military spending in our sample of countries. Time-specific shocks such as the 9/11 attacks in 2001, which led to the militarization of the MENA region and higher security risk for countries, or events such as financial crises, are also controlled by including time-fixed effects, reducing endogeneity concerns due to possible omissions of such time-invariant country- and year-specific factors.

The remainder of this paper is structured as follows: section two reviews the related literature; section three presents the data and our empirical strategy; and the results are presented and discussed in section four. Section five concludes the paper.

2   Review of the literature

Understanding the determinants of military spending is important, as it can have both positive and negative effects on economic growth. A group of studies that found a positive effect of military spending on economic growth refers to the Keynesian hypothesis. Keynesian theory focuses on military spending as a part of total aggregated demand. Assuming the existence of idle economic resources (labor and capital), higher military spending increases total demand for goods and services, leading to higher national outputs and employment. Using a panel of MENA countries, Yildirim et al. show positive growth effects of military spending from 1980–99.[6] In an earlier study, I examined the case of Iran from 1959–2007, and show that the response of income growth to positive shock in the military budget is positive and statistically significant.[7]

Other literature suggests a negative effect of higher military spending on economic growth. The main argument works through the supply-side channel, which focuses on the opportunity cost of scarce resources. Military spending diverts scarce economic resources (labor and capital) from more productive activities and spending, such as on education and health. In a case study of Iran, I show that following positive oil revenue shocks, the response of military spending is positive, while that of education and health is negative and statistically significant.[8]

Higher military spending may also increase the budget deficit and external debt. This is more problematic in oil-poor countries. Higher budget deficits following a higher military burden may also increase tax rates, discouraging private investment and reducing economic growth.[9]

[6] Yildirim, Sezgin, and Ocal, "Military Expenditure and Economic Growth in Middle Eastern Countries: A Dynamic Panel Data Analysis", *Defence and Peace Economics* 16.4 (2005), pp. 283–95.

[7] Farzanegan, "Military Spending and Economic Growth: The Case of Iran", *Defence and Peace Economics* 25.3 (2014), pp. 247–69.

[8] Farzanegan, "Oil Revenues Shocks and Government Spending Behavior in Iran", *Energy Economics* 33.6 (2011), pp. 1055–69.

[9] See, for example: Chan, "Defense Burden and Economic Growth: Unraveling the Taiwanese Enigma", *The American Political Science Review* 82.3 (1988), pp. 913–20; Lebovic and Ishaq, "Military Burden, Security Needs and Economic Growth in the Middle East", *Journal of Conflict Resolution* 31.1 (1987), pp. 106–38; Mintz and Huang, "Defense Expenditures, Economic Growth and the Peace Dividend", *American Political Science Review* 84.4 (1990), pp. 1283–93; Asseery, "Evidence from Time Series on Militarising the Economy: The Case of Iraq", *Applied Economics* 28.10 (1996), pp. 1257–61.

In addition, it is shown that military spending is accompanied by higher corruption, which in turn can throw sand in the wheels of the economy.[10]

2.1 Oil and military spending

Understanding the oil-rent-internal-conflict nexus helps shed more light on the possible effect of oil rents on military spending. How do oil rents affect the internal stability of countries? Political regimes in oil-rich states are interested in staying in power and protecting their economic rents over the long term. One mechanism is to buy the political support of potential oppositions through distribution of oil rents in the form of large-scale subsidies and provision of public jobs. Such cash transfers, which are not funded by tax revenues, can reduce the political pressure for accountability from the people by assisting the political elite in consolidating influence so long as oil rents are available. The smaller the size of the population, the higher the financial leverage of oil rents. For example, according to OPEC's annual statistics,[11] the average per capita oil revenue for Qatar from 1960–2015 was $10,994, while for Iran it was $392. This shows a significantly higher financial advantage for the Qatari government over the population in a time of crisis. If this financial leverage buys peace through (re-)distribution of rents, one should not necessarily expect a positive and significant effect of oil rents on the military category, keeping other variables constant. The response of some GCC countries to the Arab Spring events can also show this channel:

> Shortly after the collapse of former Egyptian president Hosni Mubarak's regime in February 2011, the Saudi Arabian government announced a social welfare program worth $10.7 billion to spend on new employment opportunities and loan forgiveness, a program that reached $93 billion in March 2011. Similar initiatives were introduced in the UAE, Qatar, Oman and Bahrain.[12]

Andersen and Aslaksen indicate that the type of government matters in the final effect of rents on stability of regimes.[13] Bjorvatn and Farzanegan show it is not only the type of government but also its *degree of factionalism* that matters in the final effect of rents on stability.[14] The degree of government factionalism is measured by the index of government political fractionalization (varying from 0 to 1): "the probability that two deputies picked at random from among the government parties will be of different parties".[15] Oil rents promote internal stability when the incumbent is sufficiently powerful, as is the case in most GCC countries. In these countries, the political fractionalization is rather low and there is a high degree of political imbalance.

Bjorvatn and Farzanegan present empirical evidence for their theoretical predications by using data for more than 120 countries from 1984–2009, and examine oil stability for the MENA region and find similar results.[16] They use panel data for twenty MENA countries from 2002 to 2012, and show that when regime strength is high and factional politics low, then the rents may buy

---

[10] Gupta, De Mello, and Sharan, "Corruption and Military Spending", *European Journal of Political Economy* 17.4 (2001), pp. 749–77.

[11] OPEC, "Annual Statistical Bulletin (ASB)" (2017).

[12] Bjorvatn and Farzanegan, "Resource Rents, Balance of Power", pp. 758–73.

[13] Andersen and Aslaksen, "Oil and Political Survival", *Journal of Development Economics* 100.1 (2013), pp. 89–106.

[14] Bjorvatn and Farzanegan, "Resource Rents, Balance of Power", pp. 758–73.

[15] Cruz, Keefer, and Scartascini, "Database of Political Institutions Codebook, 2015 Update (DPI 2015)", *Inter-American Development Bank* (2016).

[16] Bjorvatn and Farzanegan, "Natural-Resource Rents and Political Stability".

peace in the MENA region. Reducing oil financial leverage is a challenge for keeping the ruled people loyal to the system. In this case, the political establishment may put more investment in repressive tools by expanding the military network and its equipment. The negative correlation between oil revenues per capita (as a proxy for financial and political leverage of government over society) and the share of military spending in GDP (as a proxy for military burden) is evident in some GCC countries. For example, in the case of Saudi Arabia, this correlation is highly negative (-0.68). At times, the Saudi government experiences lower financial leverage of oil revenues (from $9,000 in 2014 to $5,000 in 2015), and we see a significant increase in military spending (from 10.7% of GDP in 2014 to 13.7% of GDP in 2015).[17] Similar patterns are seen in other years since 1988.

2.2   *Corruption and military spending*

The role of political corruption in boosting military projects and related investments has been investigated in the literature. There are a few empirical studies on the effect of corruption on military spending. Contrary to anecdotal evidence, Mauro provides rather insignificant evidence on this link.[18] Gupta et al. provide one of the few econometric analyses on this association.[19] They use four different sources of information for corruption for about 120 countries from 1985 to 1998. Their cross-country estimations show that corruption is significantly associated with higher military spending or higher arms procurement (as a share of GDP or as a share of total government spending).

The organization Transparency International has recently intensified its attention to corruption in the defense industry of countries around the world. This is reflected in their novel project, the Government Defence Anti-Corruption Index (GI). The GI index examines the existence and effectiveness of institutional and informal controls to manage the risk of corruption in defense and security institutions and of their enforcement. Based on a wide variety of sources and interviewees across seventy-seven indicators, the GI provides governments with detailed assessments of the integrity of their defense institutions. Worldwide, the organization classifies countries in six different risk groups, taking into account the situation of corruption and transparency in their military projects and institutions (i.e., A = very low, B = low, C = moderate, D = high, E = very high, and F = critical).[20] The corruption risk in the defense industry for Tunisia is "high", while it is "very high" for Iran, Jordan, Lebanon, UAE, and Saudi Arabia. For all other countries in the MENA region, including GCC countries such as Bahrain, Kuwait, Oman, and Qatar, the corruption risk is at the critical level.

According to the GI index report, those at the top of military establishments in many MENA/ GCC countries control purchasing, and are subject to little if any oversight. Individual interests in defense project decision-making are a dominant fact in many states in the MENA studied by Transparency International: "Acquisition planning — the process through which the state identifies what arms it will buy — is unclear or non-existent in every state studied."[21] In the case of

---

[17] It is less likely that this increase in Saudi military spending is due to the regional competition with Iran. Iran's military spending in GDP was 2.3% in 2014, and reached 2.5% in 2015. Israeli military burden even shows a decline for this example (from 6% to 5.4%).

[18] Mauro, "Corruption and the Composition of Government Expenditure", *Journal of Public Economics* 69.2 (1998), pp. 263–79.

[19] Gupta, de Mello, and Sharan, "Corruption and Military Spending", pp. 749–77.

[20] Full country assessments and datasets available online at Government Defence Anti-Corruption Index, https://government.defenceindex.org/.

[21] Abbas et al., "Regional Results Middle East & North Africa: Government Defence Anti-Corruption Index 2015", *Transparency International Defence and Security* (2015), p. 5.

Saudi Arabia, the report mentions that the "tactic of using defence purchases to solidify alliances" has led to the allocation of military budgets in a wasteful manner.[22] As a result, the Saudis hold large numbers of duplicative weapons systems, such as operationally similar Typhoon and F-15 fighter jets. In Kuwait, the state is struggling to train the necessary personnel for the purchased Patriot missile system.

There is a lack of meaningful legislative debate or oversight of defense acquisition in the MENA region, which increases the risk of corruption in military-related projects. Corruption in the military also has significant negative consequences for the overall security of a country. For example, in an interview in 2014, Iraqi Prime Minister Haider al-Abadi reported that the Iraqi army had been paying salaries to at least 50,000 soldiers who did not exist.[23] One of the reasons behind the collapse of different divisions of the Iraqi army confronting Islamic State terrorists was the widespread corruption in the military, reflected in the high numbers of "ghost soldiers".

The following sections examine the hypothesis that the effect of oil rents on military spending depends on the level of political (or administrative) corruption and its development over time across countries.

3 Research design

3.1 Data, specification, and empirical strategy

Our main hypothesis is that higher levels of political and administrative corruption increase the final effects of oil rents on military budgets. In other words, oil rents are more likely to be allocated to the military when the extent of corruption is relatively high. In this context, we also control for other variables that may influence the military-oil-rents nexus (e.g., economic development, size of population, risk of internal and external conflict, trade, total government spending, education level of population, rule of law, and quality of democratic institutions). This strategy helps reduce the risk of ignoring other important determinants of military spending noted in the literature.

We test our hypothesis by using panel regressions for five GCC countries from 1984–2014. We also compare the results of the GCC sample to our analysis of non-GCC countries in the MENA region. To estimate whether the relationship between oil rents and military spending varies systematically with the level of corruption, we use the following specification:

$$military_{it} = \alpha + \beta_1 \cdot oil_{it-4} + \beta_2 \cdot corruption_{it-4} + \beta_3 \cdot (oil_{it-4} \times corruption_{it-4})$$
$$+ \beta_4 \cdot Z_{it-1} + u_i + \theta_t + \varepsilon_{it},$$

with country i and time t, where $military$ is for military spending, oil is a measure of oil rents dependency, $corruption$ is a measure of the perception of corruption, $oil \times corruption$ is the interaction of oil and corruption, and Z is the control variables. To reduce the possible reverse feedback, we use a four-year lag of key explanatory variables such as oil, corruption, and their interaction. All other explanatory variables are lagged at one year. According to our expectations, the sign of the interaction term coefficient should be positive ($\beta_3 > 0$); the higher the level of corruption, the higher the effect of oil rents on military spending.

---

[22] Ibid., p. 6.
[23] Morris, "Investigation Finds 50,000 'Ghost' Soldiers in Iraqi Army, Prime Minister Says", *The Washington Post*, 30 Nov. 2014.

The marginal effect of oil rents on military spending can be calculated by examining the following partial derivative:

$$\frac{\partial(military_{it})}{\partial(oil_{it-4})} = \beta_1 + \beta_3 \cdot (corruption_{it-4})$$

3.1.1 Dependent variable: military spending

Our dependent variable is a measure of military spending. In this study, we use the logarithm of military expenditures per capita from the World Bank. The World Bank reports military expenditure data from the Stockholm International Peace Research Institute (SIPRI), which is derived from the NATO definition. The military spending data covers all current and capital expenditures on the armed forces including peacekeeping forces; defense ministries and other government agencies engaged in defense projects; paramilitary forces — if these are judged to be trained and equipped for military operations; and military space activities. These expenditures cover military and civil personnel, including retirement pensions of military personnel and social services for personnel; operation and maintenance; procurement; military research and development; and military aid (in the military expenditures of the donor country). The reported military spending is not including civil defense and current expenditures for previous military activities, such as for veterans' benefits, demobilization, conversion, and destruction of weapons. In our sample of GCC countries, the lowest per capita military spending is observed for Bahrain ($167) in 1988, and the highest is recorded for Saudi Arabia in 2015 ($10,312).

3.1.2 Independent variables

3.1.2.1 Oil rents

Our main focus is the effect of resource rents on military spending. In the GCC region, which is our (main) sample of analysis, the major natural resource wealth is crude oil. In this study, following Atkinson and Hamilton,[24] Sachs and Warner,[25] and Bjorvatn and Farzanegan,[26] to measure the relative importance of oil in the economies of the region, we use the share of oil rents as a percentage of GDP. Oil rents are the difference between the value of crude oil production at world prices and total costs of production. This leads to a more realistic picture of oil wealth than that of oil export revenues. In the GCC sample, the range of oil rents dependency varies from 1.7% (Bahrain in 1998) to 60% of GDP (Kuwait in 2011). The average level of oil rents dependency in GCC countries in our study is approximately 25%. Within the larger MENA region, Israel, Jordan, Lebanon, Malta, Morocco, and the West Bank have oil rents as a share of GDP at less than 1%. There are also countries in which oil rents comprise more than half of their economies, such as Iraq, Kuwait, Libya, and Saudi Arabia. We extract oil rents (percent of GDP) from the World Bank. Oil rent estimates are based on sources and methods explained

---

[24] Atkinson and Hamilton, "Savings, Growth and the Resource Curse Hypothesis", *World Development* 31.11 (2003), pp. 1793–807.

[25] Sachs and Warner, "The Curse of Natural Resources", *European Economic Review* 45.4–6 (2001), pp. 827–38.

[26] Bjorvatn and Farzanegan, "Demographic Transition in Resource Rich Countries: A Bonus or a Curse?", *World Development* 45 (2013), pp. 337–51.

in "The Changing Wealth of Nations: Measuring Sustainable Development in the New Millennium".[27]

3.1.2.2 Corruption

Another key explanatory variable is the level of corruption. There are different definitions of corruption. We follow a larger part of the related literature, and define corruption as abuse of public office for private benefit.[28] Misuse of state resources to expand political power refers to political corruption, which is a relevant factor in shaping the effects of oil rents on the military.[29] For our analysis, we follow corruption measurements from the International Country Risk Guide (ICRG).[30] The ICRG corruption index is from 0 to 6 (most to least corrupt); we have re-scaled the index from 1 to 7 (least to most corrupt).

The main focus of the ICRG corruption index is on the public sector. For this purpose, the ICRG corruption index covers "actual or potential corruption in the form of excessive patronage, nepotism, job reservations, 'favor-for-favors', secret party funding, and suspiciously close ties between politics and business".[31] Of course, the ICRG corruption index does not neglect corruption at the private sector. They consider "financial corruption in the form of demands for special payments and bribes connected with import and export licenses, exchange controls, tax assessments, police protection, or loans".[32]

One of the advantages of the ICRG corruption index is its long-term coverage, since 1984. Another reason for using the index is its consistency in measuring corruption over time and across countries. Given the fact that ICRG corruption is not a composite index, its year-to-year comparisons are more reliable than other indicators, such as Transparency International and World Governance Indicators (WGI), which are composite indicators and their underlying data or weighting methods may change over time.[33]

Almost all corruption indicators such as ICRG, WGI, and Transparency International, measure the perception of corruption, and are based on country experts, business owners, NGOs, and household opinion. There are pro and contra debates on the validity of subjective indicators of corruption. On one hand, studies such as Fisman and Miguel,[34] and Fisman and Wei,[35] present evidence on the objective validity of the corruption perception indicators. On the other, studies such as Olken,[36] and Donchev and Ujhelyi,[37] undermine a significant link between the

---

[27] World Bank, "The Changing Wealth of Nations: Measuring Sustainable Development in the New Millennium" (2011).

[28] Shleifer and Vishny, "Corruption", *Quarterly Journal of Economics* 108.3 (1993), pp. 599–617.

[29] Rose-Ackerman, *Corruption and Government: Causes, Consequences and Reform* (1999); Manzetti and Wilson, "Why Do Corrupt Governments Maintain Public Support?", *Comparative Political Studies* 40.8 (2007), pp. 949–70.

[30] Howell, "International Country Risk Guide Methodology", *The PRS Group* (2015).

[31] Ibid.

[32] Ibid.

[33] Hessami, "Political Corruption, Public Procurement, and Budget Composition: Theory and Evidence from OECD Countries", *European Journal of Political Economy* 34 (2014), pp. 372–89.

[34] Fisman and Miguel, "Corruption, Norms, and Legal Enforcement: Evidence from Diplomatic Parking Tickets", *Journal of Political Economy* 115.6 (2007), pp. 1020–48.

[35] Fisman and Wei, "The Smuggling of Art, and the Art of Smuggling: Uncovering the Illicit Trade in Cultural Property and Antiques", *American Economic Journal: Applied Economics* 1.3 (2009), pp. 82–96.

[36] Olken, "Corruption Perceptions vs. Corruption Reality", *Journal of Public Economics* 93.7–8 (2009), pp. 950–64.

[37] Donchev and Ujhelyi, "What Do Corruption Indices Measure?", *Economics & Politics* 26.2 (2014), pp. 309–31.

perception indicators of corruption and the real experience of corruption. Nevertheless, these perception-based indicators of corruption are widely used in the related literature.[38] Treisman also shows that all of the corruption perception indicators are highly correlated,[39] indicating the fact that they are measuring a similar concept.

Another concern about the perception-based indicators is related to the level of information flow and press freedom across countries. Countries with higher press censorship may restrict the flow of information about corruption scandals, and thus the perception of the real extent of corruption may be distorted. This concern has been investigated in the literature, and studies such as Besley and Prat,[40] Brunetti and Weder,[41] and Sung,[42] show that corruption perception is even higher in countries with lower press freedoms. In the GCC sample, the rescaled ICRG corruption index has a minimum of 3 (Bahrain in the early 1990s) to a maximum of 5 (Bahrain in 2000s; Kuwait in early 2000s; Qatar, Saudi Arabia, and UAE for most years of analysis). In the larger sample of the MENA, the rescaled ICRG corruption index has a minimum of 2 (Israel) and maximum of 6 (Iraq, Lebanon, Libya) on a 1–7 scale.

3.1.2.3   *Country and time-fixed effects*

In addition, there are some country-specific characteristics that can be important for military spending. These country-specific factors are time-invariant. For example, geographical location of countries does not usually change over time, and, at the same time, they are relevant for the stability of countries and the budget amount they allocate to the military. Some countries in geostrategic locations may attract attention and interventions of external powers, and in return shape the military budget of the affected countries. US military bases located in some GCC countries can be also related to country-specific factors and relevant for the host country's military spending.

Another country-specific characteristic that may affect military budgets is ethnic fractionalization. The fractionalization of countries may have implications for the stability of political systems. In particular, it is shown that ethnic fractionalization increases the conflict risk of resource rents.[43] According to the fractionalization dataset,[44] which measures the degree of ethnic, linguistic, and religious heterogeneity in various countries, we can observe a significant cross-country variation in the GCC region. In some GCC countries, such as UAE and Qatar, due to a significant share of migrant populations from different parts of world, we see a higher degree of ethnic fractionalization in related datasets. For example, ethnic fractionalization (in a range of 0 to 1) is lowest in Saudi Arabia (0.18); followed by Oman (0.43); Bahrain (0.52); UAE (0.62); Kuwait (0.66); and Qatar (0.74). In terms of religious fractionalization, the picture is different: the lowest is calculated for Qatar (0.09), and the highest for Kuwait (0.67).

---

[38] Farzanegan and Witthuhn, "Corruption and Political Stability: Does the Youth Bulge Matter?" *European Journal of Political Economy* 49 (2017), pp. 47–70.

[39] Treisman, "The Causes of Corruption: A Cross-national Study", *Journal of Public Economics* 76.3 (2000), pp. 399–457.

[40] Besley and Prat, "Handcuffs for the Grabbing Hand? Media Capture and Government Accountability", *American Economic Review* 96.3 (2006), pp. 720–36.

[41] Brunetti and Weder, "A Free Press is Bad News for Corruption", *Journal of Public Economics* 87.7–8 (2003), pp. 1801–24.

[42] Sung, "A Convergence Approach to the Analysis of Political Corruption: A Cross-national Study", *Crime, Law and Social Change* 38.2 (2002), pp. 137–60.

[43] Collier and Hoeffler, "On Economic Causes of Civil War", *Oxford Economic Papers* 50.4 (1998), pp. 563–73.

[44] Alesina et al., "Fractionalization", *Journal of Economic Growth* 8.2 (2003), pp. 155–94.

Regarding language fractionalization, the lowest score is recorded for Saudi Arabia (0.09), and the highest score is for UAE (0.48). Such time-invariant factors are controlled in our estimations by country-fixed effects.

Likewise, there are time-specific factors that can shape the military budgets of countries. A specific event that occurs in one year, such as on 11 September 2001, may increase the conflict risk around the world, affecting the military spending of many countries in our sample. Another example is the Arab Spring, which shaped the security risk of many countries in the MENA region and their military spending. Oil price shock or financial crisis in a specific year may affect the military budgets of many countries in our sample at the same time. Such time-specific factors are controlled in our analysis through time-fixed effects.

3.1.3 Control variables

Besides the main variables of interest, such as oil rents, corruption, and their interaction term, we also control for another set of other drivers of military spending in our estimations. We follow Gupta et al. in the selection of control variables.[45] The source for the control variables (with the exception offset of ICRG political risk related variables) is the World Bank. In the following, we briefly explain their association with military spending.

3.1.3.1 Economic factors

One of the main control variables is the GDP per capita. We use GDP per capita (constant 2010 in US dollars) as a proxy for scale of economy. Increasing economic development can foster education and public awareness as to the importance of other critical socioeconomic and environmental concerns, requiring the state to allocate more of the budget to public goods and services. In addition, economic development is correlated with the dependent variable (military spending) and the key explanatory variables (oil rents dependency and corruption). In order to isolate the effect of the key explanatory variables on military spending, different stages of economic development should be taken into account.

The ratio of government spending to GDP is another control variable that is often used as a predicator of military spending in structural models of military spending.[46] We expect to see a positive association between total government spending and military spending. A channel through which higher income per capita may affect military spending is also through total government spending. Although some authors such as Dunne et al. show a negative effect of income per capita on military burden,[47] others such as Collier and Hoeffler,[48] show a positive effect. Their argument is that higher economic development increases the capacity of the state to raise taxes and borrow, facilitating the funding of government spending, including on the military. The average share of government spending in GDP of the GCC countries is 21%, varying from approximately 7% (UAE in 2006 and 2007) to 55% (Kuwait in 1992).

Trade as a share of GDP is controlling for the degree of integration of a country in international markets, a measure of trade openness. The higher level of economic globalization following higher intensity of trade in the economy may foster growth and political stability. As a result, the risk of

---

[45] Gupta, de Mello, and Sharan, "Corruption and Military Spending", pp. 749–77.

[46] Hewitt, "Military Expenditures Worldwide: Determinants and Trends, 1972–88", *Journal of Public Policy* 12.2 (1992), pp. 105–52.

[47] Dunne, Perlo-Freeman, and Smith, "The Demand for Military Expenditure in Developing Countries: Hostility Versus Capability", pp. 293–302.

[48] Collier and Hoeffler, "Military Spending and the Risks of Coups d'etat".

external and internal conflicts may be reduced. The lower instability risk may also reduce the necessity of spending on military establishment. In 1758, Montesquieu also reflected on the positive role of trade in reducing conflict and thus military spending: "The natural effect of trade is to bring about peace. Two nations which trade together render themselves reciprocally dependent."[49] However, higher intensity of trade may increase the competition in the market and cause significant adjustments in protected industries. This may cost jobs especially for a lower-skilled labor force. If trade globalization increases the income gap between the rich and the poor, then the risk of conflict can also increase, pushing the necessity of more spending on security forces. We measure trade openness as a sum of total exports and imports divided by GDP. Since most MENA countries have significant oil exports, they may show higher intensity of trade. We also test the results by using share of total imports in GDP. The average share of trade in GDP of the GCC region is 103%, ranging from 56% (Saudi Arabia in 1998) to 210% (Bahrain in 1990).

3.1.3.2   Socio-demographic factors

Among socio-demographic factors, we control for secondary school enrollment rate, which indicates the country level of social and human development. This is the ratio of total enrollment regardless of age, to the population of the age group that officially corresponds to the education level. Secondary education provides the foundations for long-term learning and human development by offering more subject- or skill-oriented instruction using teachers who are more specialized. Educated people tend to interact with others due to their communication abilities gained through education. Hence, they are more likely to participate in public debates. The educated population is also more aware of their legal rights, such as voting as a means to demand political reforms, which might restrict military elites and their interest groups. This might also increase the possibility of reducing the share of military spending. The empirical evidence backs the theoretical arguments across countries.[50] Within the GCC region, the average secondary school enrollment rate is 84.6%, varying from 21.2% (Oman in 1984) to 111% (Qatar in 2005).

We also consider in our estimates the size of population that controls for the available human capital for defending the security of a country. Highly populated countries may build up more labor-intensive military units. For example, the Arab countries of the Persian Gulf region, with relatively smaller populations compared with Iran, have more capital-intensive military organizations and are more dependent on imported arms. A greater population size can also create a natural sense of security against external threats, lowering the need for capital-intensive military projects. The negative association between population and military spending is also observed in other studies.[51] Dunne and Perlo-Freeman offer two explanations for this negative link: first, a larger population offers security in itself; and second, a larger population may make the need for state civil spending more of a priority than security needs.[52] We expect a negative association

---

[49] Martin, Mayer, and Thoenig, "Does Globalisation Pacify International Relations?" *Vox* (2007).

[50] Barro, "Determinants of Democracy", *Journal of Political Economy* 107.6 (1999), pp. 158–83; Glaeser, Ponzetto, and Shleifer, "Why Does Democracy Need Education?", working paper submitted to the *National Bureau of Economic Research* 12128, Cambridge (2006); Castello-Climent, "On the Distribution of Education and Democracy", *Journal of Development Economics* 87.2 (2008), pp. 179–90.

[51] Dunne and Perlo-Freeman, "The Demand for Military Spending in Developing Countries: A Dynamic Panel Analysis", *Defence and Peace Economics* 14.6 (2003), pp. 461–74; Dunne, Perlo-Freeman, and Smith, "The Demand for Military Expenditure in Developing Countries: Hostility Versus Capability", *Defence and Peace Economics* 19.4 (2008), pp. 293–302; Collier and Hoeffler, "Military Spending" (2007).

[52] Dunne and Perlo-Freeman, "The Demand for Military Spending in Developing Countries", *International Review of Applied Economics* 17.1 (2003), pp. 23–48.

between size of population and military spending. In the GCC, countries such as Bahrain, Kuwait, Oman, and Qatar have population sizes of less than five million people. Saudi Arabia is the most populous country in the GCC with more than thirty million people.

3.1.3.3 Institutional factors

Besides corruption, which is a key moderating channel in the military-spending-oil-rents nexus in our study, we control for other institutional dimensions; one being law and order. The ICRG explains that the "law" element considers the strength and impartiality of the legal system, while "order" is an assessment of popular observance of the law. It ranges from 0 to 6; the higher scores reflect a better situation in rule of law and order in a country. Strong checks and balances may increase the quality of government spending and the transparency of oil rents allocation in government budgets. It also increases the costs of corruption, hindering the shift of oil rents to capital-intensive military projects, which are more attractive for bribe-seekers.

ICRG also provides country scores on their internal and external conflict risks. Such conflict risks can increase the necessity of military spending. The internal conflict index of ICRG is an assessment of political violence within a country and its actual or potential impact on governance. It ranges from 0 to 12, with the highest score for those countries with no armed or civil opposition to the government. In these highly internally stable countries, the government does not indulge in arbitrary violence — direct or indirect — against its own people. The lowest rating is given to a country embroiled in an ongoing civil war. There are three different sub-components in the internal conflict index, including civil war/coup threat, terrorism/political violence, and civil disorder. The external conflict index has a similar range (0 to 12). It is an assessment of the risk to the incumbent government from foreign action. These risks can be in the form of nonviolent external pressure (diplomatic pressures, withholding of aid, trade restrictions, territorial disputes, sanctions, etc.) and/or violent external pressure (cross-border conflicts to all-out war). The highest score is given to countries with very low risk. The three subcomponents of this index are war, cross-border conflicts, and foreign pressures. We also consider the ethnic tension index of the ICRG. It is an assessment of the degree of tension within a country attributable to racial, national, or linguistic divisions, and it ranges from 0 to 6. Lower ratings are given to countries where racial and nationality tensions are high. Such tensions may be caused by a lack of social-political tolerance or inclusive growth policy in a country. Higher scores are for countries with tensions that are low.

The quality of political institutions is also an important factor in the allocation of oil rents to military and nonmilitary categories of spending.[53] Countries that are more democratic consider the needs and priorities of a larger cohort of the population in their spending behavior, while autocratic regimes rely more on military elites. We use the ICRG democratic accountability index for this purpose. It measures from 0 to 6, with the higher number showing a better quality political institution. This index is a measure of how responsive government is to its people. Lower accountability increases the risk of political collapse, either peacefully (in democracies) or violently (in non-democracies).

3.1.3.4 External factors: Military spending of border countries

Among the external factors relevant for our analysis, we consider the average military spending of neighboring countries. Countries may adjust their military budget, taking into account the

---

[53] Dizaji, Farzanegan, and Naghavi, "Political Institutions and Government Spending Behavior: Theory and Evidence from Iran", *International Tax and Public Finance* 23.3 (2016), pp. 522–49.

spending behavior of their immediate neighbors. The concept of a "security web" was introduced by Rosh,[54] as a response to the shortcomings of the "arms race model" in explaining the drivers of military spending.[55] To control for regional arms competition and regional tension, which may influence a country's military spending, we use the unweighted average of neighboring countries' ratio of military spending to GDP.[56] Dunne and Perlo-Freeman also suggest that determinants of military spending changed after the end of the Cold War.[57] Since the 1990s, we have been observing a larger share of internal conflicts that highlight the importance of domestic socioeconomic, demographic, and institutional factors as determinants of military spending. To measure the statistical accuracy and reliability of our estimated coefficients, we use robust standard errors.[58]

4   Main results

Table II presents the country and year fixed-effects regression results, which show how "within-country" changes in the explanatory variables, such as oil rents and corruption, affect the within-country changes in military spending in the sample of the GCC region.

In line with our theoretical expectation, the positive interaction term between oil rents and corruption is robust in its signs of impacts, size, and significance in 11 out of 12 models. The final effect of oil rents on military spending per capita depends on the level of political and administrative corruption. At higher levels of corruption, more oil rents are channeled more significantly to military projects. The direct effect of oil rents in all models is negative on military spending per capita and is statistically significant. The within-country increases in oil rents in the GCC states reduces the per capita military spending within the countries of this region. Using panel regression for more than 120 countries from 1984 to 2009, Bjorvatn and Farzanegan show that oil rents may buy political stability in political systems with a low degree of political fractionalization; i.e., when political power is sufficiently concentrated.[59] Oil rents can intensify rent-seeking and risk of conflict if the political system is factional; i.e., when the incumbent is sufficiently weak. Bjorvatn and Farzanegan give an example of the reaction of GCC rulers to the Arab Spring events to illustrate how cooperation funded by oil rents may stabilize the political system, and thus reduce the necessity for higher military spending.

According to their model, the power of a regime relative to the opposition shapes the costs of co-opting opposition groups. A powerful incumbent does have a lower price tag for stability compared to a powerless one when oil rents increase. Therefore, Bjorvatn and Farzanegan argue that balance of power in society is important for the final stability effects of oil rents. The political stability in their analysis is the assessment of political violence in the country and its actual or potential impact on governance from the ICRG dataset.

In another related study, Bjorvatn and Farzanegan examine the political-stability-oil-rents nexus and the role of balance of power within the twenty MENA countries from 2002 to

---

[54] Rosh, "Third World Militarization: Security Webs and the States They Ensnare", *Journal of Conflict Resolution* 32.4 (1988), pp. 671–98.

[55] Majeski and Jones, "Arms Race Modeling: Causality Analysis and Model Specification", *Journal of Conflict Resolution* 25.2 (1981), pp. 259–88.

[56] Davoodi et al., "Military Spending, the Peace Dividend, and Fiscal Adjustment", working paper submitted to the *International Monetary Fund (IMF)* 99.87, Washington, DC (1999); Gupta, de Mello, and Sharan, "Corruption and Military Spending", pp. 749–77.

[57] Dunne and Perlo-Freeman, "The Demand for Military Spending" (2003), pp. 23–48.

[58] Wooldridge, *Econometric Analysis of Cross Section and Panel Data* (2002).

[59] Bjorvatn and Farzanegan, "Resource Rents, Balance of Power", pp. 758–73.

2012. They show that oil rents increase stability within the MENA region but political fractiona-lization reduces this positive effect. In their words, "rents are stabilizers when the regime strength is high and factional politics is low, and works as a destabilizing force in regimes that are weak from the outset".[60]

Another interesting observation in Table II is the effect of corruption on military spending per capita. Within the oil-rich GCC countries, the effect is negative and statistically significant in seven models (6–12). Increasing political corruption within the GCC countries seems to reduce military spending. This initially surprising effect becomes clearer if we focus on the stability effects of corruption, especially within the oil-rich states. Fjelde and Hegre argue that political corruption in non-democratic systems allows leaders to build political support, and increases the duration of their power.[61] In another related study, Fjelde shows in her world-wide sample that corruption can moderate the negative effect of oil rents on political stability.[62] Her analysis concludes that "selective accommodation of private interests", which can happen more easily under corrupt systems, can dampen the conflict risk of increasing oil rents. These findings can then imply lower necessity for military spending in oil-rich and corrupt economies.

Our main hypothesis implies that there is an interaction between oil rents and corruption. Increasing oil rents may lead to higher spending on the military when corruption is also increas-ing. There is a positive interaction term in all models in Table II. The estimated interaction coeffi-cient is also statistically significant in 11 of 12 models. Note that we are using the same sample of observation in all twelve models. The changes in size of effects, therefore, is not due to different observations across models.

We have examined the robustness of our main finding (the interaction of oil and corruption) by including a different set of control variables in models 1 to 12. Among control variables, we see the negative effect of economic development (real GDP per capita) on military spending. The negative effect is in line with our previous expectation that higher levels of development increase the economic perspective, employment, and investment, among others. Because of development, countries experience a higher political stability, reducing the willingness to increase the size of military forces. The negative impact of income per capita on military spending per capita within the GCC sample is, however, not statistically significant.

The effect of secondary education on the military budget is only marginally statistically sig-nificant. As in previous literature, we also find a highly significant and positive effect of total government spending (percent of GDP) on military spending in the GCC sample in models 3–6. Total government spending includes both military and nonmilitary categories such as education and health. The significant positive effect of government total spending (as a percent of GDP) in GCC countries on military spending sheds more light on the political economy of GCC govern-ments' fiscal behavior and the leverage of military elites in budget allocations in this region. The negative effect of population size on military spending is robust and statistically significant in all models. This is comparable to the rest of the literature. Larger populations may increase pressure on the provision of public goods, such as education and health, reducing the size of the military category from the overall budget. The higher share of trade in GDP in the GCC countries shows a marginal dampening effect on military spending. As I have discussed

---

[60] Bjorvatn and Farzanegan, "Natural-Resource Rents and Political Stability", pp. 36.

[61] Fjelde and Hegre, "Political Corruption and Institutional Stability", *Studies in Comparative International Development* 49.3 (2014), pp. 267–99.

[62] Fjelde, "Buying Peace? Oil Wealth, Corruption and Civil War, 1985–99", *Journal of Peace Research* 46.2 (2009), pp. 199–218.

Table II: Per capita military spending, oil rents, and corruption (country and time-fixed effects OLS panel regressions). Sample of GCC countries with four years lag of oil, corruption, and their interactions

	(1)	(2)	(3)	(4)	(5)	(6)	(7)	(8)	(9)	(10)	(11)	(12)
	Dep. Variable: log per capita military spending											
Oil	-0.685*	-1.513***	-1.291***	-1.231***	-1.142**	-1.209***	-1.331***	-1.364***	-1.307***	-1.301***	-1.259***	-1.175***
	(-2.56)	(-4.61)	(-5.40)	(-6.45)	(-4.54)	(-6.39)	(-8.71)	(-9.56)	(-6.44)	(-5.70)	(-4.98)	(-4.09)
corruption	-0.094	-0.180	-0.183	-0.154	-0.114	-0.178**	-0.223**	-0.239**	-0.222**	-0.224*	-0.248*	-0.315*
	(-0.41)	(-0.94)	(-1.38)	(-2.08)	(-1.66)	(-2.95)	(-3.52)	(-3.59)	(-2.87)	(-2.73)	(-2.58)	(-2.52)
*oil*corruption*	0.108	0.136*	0.137**	0.095***	0.101***	0.094***	0.111***	0.113**	0.109**	0.103***	0.125***	0.144**
	(1.16)	(2.44)	(2.79)	(6.97)	(7.25)	(6.14)	(5.30)	(4.27)	(3.54)	(3.45)	(5.31)	(4.12)
GDP per capita	0.254	-0.557	-0.068	-0.053	-0.232	0.043	-0.111	-0.095	-0.037	-0.007	-0.000	0.224
	(0.42)	(-0.99)	(-0.18)	(-0.17)	(-0.96)	(0.09)	(-0.21)	(-0.17)	(-0.06)	(-0.01)	(-0.00)	(0.40)
secondary education		1.131*	0.618*	0.516**	0.645*	0.430	0.487	0.478	0.435	0.403	0.407	0.082
		(2.73)	(2.17)	(2.20)	(2.53)	(1.44)	(1.90)	(1.90)	(1.48)	(1.12)	(1.28)	(0.17)
government spending			0.899**	0.815***	0.945***	0.718**	0.591	0.498	0.497	0.473	0.504	0.375
			(3.75)	(5.07)	(6.18)	(2.87)	(1.69)	(1.24)	(1.31)	(1.11)	(1.20)	(0.79)
population				-0.591**	-0.604**	-0.608***	-0.521**	-0.482*	-0.497*	-0.518**	-0.460*	-0.410*
				(-4.06)	(-4.58)	(-4.63)	(-2.81)	(-2.30)	(-2.33)	(-2.86)	(-2.48)	(-2.32)
trade openness					-0.514*							
					(-2.43)							
import openness						0.142	0.174	0.250	0.230	0.243	0.209	0.290
						(0.48)	(0.52)	(0.68)	(0.64)	(0.64)	(0.63)	(0.72)
rule of law							0.077	0.138	0.124	0.130	0.142	0.177
							(1.01)	(1.17)	(0.97)	(0.94)	(0.96)	(1.26)
internal stability								-0.021	-0.014	-0.006	-0.017	0.003
								(-1.12)	(-0.63)	(-0.34)	(-0.98)	(0.20)
external stability									-0.015	-0.015	-0.028	-0.028
									(-0.87)	(-0.80)	(-0.96)	(-0.73)
lack of ethnical conflict										-0.036	-0.025	-0.098
										(-0.46)	(-0.32)	(-1.05)
democracy											0.040	0.045
											(1.10)	(0.90)
Ave. military spending of neighbors												0.732
												(1.40)
Obs.	72	72	72	72	72	72	72	72	72	72	72	72
R-sq.	0.66	0.81	0.86	0.89	0.90	0.89	0.90	0.90	0.90	0.90	0.90	0.91

Robust *t-statistics* are in parentheses (clustered standard errors at country level). All independent variables are lagged by one year. ***, **, and * indicate significance at the 1%, 5%, and 10% levels, respectively. All variables (except for ICRG indexes) are in logarithmic form. Country and time-fixed effects are included in all models.

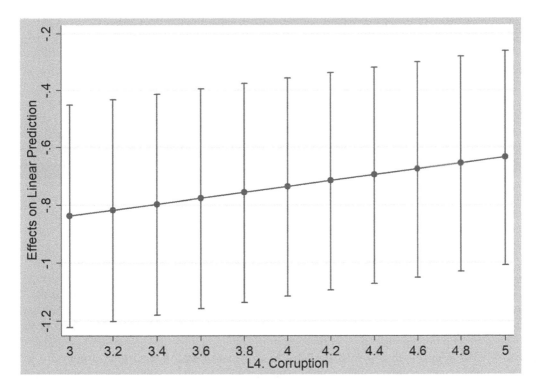

Figure 1. Marginal effect of oil rents on military spending per capita at different levels of corruption in GCC countries.
Note: the marginal effects are based on Model 5 in Table 2. The solid middle line shows the marginal effects, and dashed lines are confidence intervals at 90 percent level. The five countries included in this model are Bahrain, Kuwait, Qatar, Oman, and Saudi Arabia. L4 stands for a four-year lag of variable.

previously,[63] the effectiveness of trade openness on developmental outcomes depends on the quality of political institutions. Higher transparency and the free flow of information as characteristics of a democratic regime may shed more light on the quality of imported goods (military versus nonmilitary), especially in the GCC region, with imports funded mostly through oil exports.

Besides the abovementioned factors, we have not found any other significant institutional and/ or regional determinants of military spending per capita in the GCC states. According to the R-squared values, models 4–12 explain 90% of within-country variation of military spending per capita in our GCC sample of countries. Using Eq. 2, we can also estimate and illustrate the marginal effect of an increase in oil rents on military spending per capita at different levels of corruption in the GCC sample. Figure 1 shows the marginal effects and their associated 90% confidence intervals.

We observe that when corruption is at its lowest level (score of 3 in a range of 1–7 in the GCC sample), then an increase in oil rents has the most negative and significant effect on

[63] Farzanegan, "Illegal Trade in the Iranian Economy: Evidence from Structural Equation Model", *European Journal of Political Economy* 25.4 (2009), pp. 489–507.

military spending. By increasing corruption, we observe that the negative effect of oil rents on military spending is reduced. In other words, oil rents may increase military spending at higher levels of political corruption in the GCC sample. The overall effect of oil rents on military spending, however, remains in the negative zone, supporting the arguments of Bjorvatn and Farzanegan.[64]

We also estimated our model for the sample of non-GCC countries in the MENA region. The results are shown in Table III. There are similarities and differences with the GCC sample. The main similarity is the sign and significance of the joint effect of oil rents and corruption on military spending per capita. The interaction term is positive and statistically significant in all twelve models. The main differences are related to the direct effects of oil rents and corruption on military spending in the non-GCC sample of the MENA region. One of them is the lack of direct negative effect of oil rents on military spending. The direct effect of oil rents is statistically insignificant. In addition, the direct effect of corruption in this sample on military spending is positive and statistically significant in models 1–8. This is in line with earlier literature. For example, Gupta et al. also find a similar positive effect of corruption on military in their cross-country and panel data regressions for 120 countries during 1985–98.[65] There are many examples of bribes related to government spending on the military. According to Tanzi, approximately 15% of total spending on arms is estimated to be related to payments of bribes.[66] Transparency International estimates that at least $20 billion is lost to corruption in the military sector every year worldwide. Our results can also be read as a double burden of corruption in oil-rich countries of the non-GCC countries in the MENA region. More corruption leads to more spending on the military not only in these countries, but such misallocation of government budgets is higher in countries with more oil rents.

We observe a significant and negative effect of higher income per capita on military spending per capita. Another main difference is the robust and negative effect of external risk of conflict (higher scores mean lower risk of external conflict) on military spending.

Figure 2 illustrates the marginal effect of oil rents on military spending per capita at different levels of corruption in the non-GCC sample of the MENA region. The quality of the effect is similar to the GCC sample: oil rents at higher levels of political corruption lead to more military spending per capita. However, for the non-GCC sample, we can see that the overall effect of oil rents is positive at all levels of corruption on military spending. This is again in line with studies of Bjorvatn and Farzanegan, in which oil rents in politically factional countries lead to more instability and thus more spending on military.[67] The factional politics is more often a characteristic of many non-GCC countries sampled in the MENA region. We are 90% confident that higher oil rents can increase military spending when the corruption is more than 2.6 (out of 7). At 95% confidence interval the critical corruption level for our sample is 3.6. In 2014, these were countries such as *Algeria*, Egypt, *Iran*, *Iraq*, Jordan, Lebanon, *Libya*, Morocco, Syria, Tunisia, and Yemen (oil-rich states are in italics).

---

[64] Bjorvatn and Farzanegan, "Resource Rents, Balance of Power", pp. 758–73; Bjorvatn and Farzanegan, "Natural-Resource Rents and Political Stability", pp. 33–7.

[65] Gupta, de Mello, and Sharan, "Corruption and Military Spending", pp. 749–77.

[66] Tanzi, "Corruption around the World: Causes, Consequences, Scope, and Cures", *IMF Staff Papers* 45.4 (1998): 559–94.

[67] Bjorvatn and Farzanegan, "Resource Rents, Balance of Power", pp. 758–73; Bjorvatn and Farzanegan, "Natural-Resource Rents and Political Stability", pp. 33–7.

Table III: Per capita military spending, oil rents, and corruption (country and time-fixed effects OLS panel regressions). Sample of non-GCC countries with four years lag of oil, corruption, and their interactions

	(1)	(2)	(3)	(4)	(5)	(6)	(7)	(8)	(9)	(10)	(11)	(12)
	Dep. Variable: log per capita military spending											
oil	0.227	0.226	0.232	0.211	0.197	0.213	0.203	0.214	0.194	0.198	0.197	0.101
	(1.30)	(1.21)	(1.23)	(1.22)	(1.20)	(1.21)	(1.09)	(1.06)	(1.07)	(1.07)	(1.05)	(0.74)
corruption	0.237**	0.238**	0.255**	0.244**	0.247**	0.218**	0.217**	0.223**	0.148	0.144	0.139	0.146
	(2.39)	(2.36)	(2.77)	(2.31)	(2.31)	(2.73)	(2.60)	(2.79)	(1.46)	(1.34)	(1.37)	(1.49)
oil*corruption	0.047**	0.047**	0.046**	0.043**	0.042**	0.044**	0.040**	0.038**	0.046***	0.046**	0.045**	0.053***
	(2.80)	(2.80)	(2.65)	(2.87)	(2.73)	(2.88)	(2.84)	(2.60)	(3.40)	(3.17)	(3.20)	(5.35)
GDP per capita	-1.584*	-1.596*	-1.545*	-1.771**	-1.688**	-1.821**	-1.602**	-1.651**	-1.425**	-1.447**	-1.520**	-1.054**
	(-2.16)	(-2.18)	(-2.16)	(-2.60)	(-2.90)	(-2.48)	(-2.41)	(-2.57)	(-2.50)	(-2.56)	(-3.03)	(-2.59)
secondary education		0.024	-0.064	-0.283	-0.353	-0.122	-0.032	-0.061	0.189	0.158	0.176	0.645
		(0.04)	(-0.12)	(-0.45)	(-0.54)	(-0.17)	(-0.05)	(-0.10)	(0.39)	(0.36)	(0.35)	(1.41)
government spending			0.276	0.134	0.180	0.254	0.337	0.394	0.446	0.464	0.403	0.339
			(0.67)	(0.31)	(0.39)	(0.58)	(0.73)	(0.93)	(0.89)	(0.92)	(0.90)	(1.24)
population				-1.345	-1.095	-1.472	-1.495	-1.634	-0.252	-0.308	-0.291	-1.482
				(-0.78)	(-0.69)	(-0.90)	(-1.09)	(-1.11)	(-0.22)	(-0.28)	(-0.26)	(-1.60)
trade openness					0.327							
					(0.67)							
import openness						-0.452	-0.453	-0.419	-0.218	-0.237	-0.180	0.092
						(-0.79)	(-0.81)	(-0.81)	(-0.51)	(-0.57)	(-0.52)	(0.26)
rule of law							-0.152	-0.181*	-0.131	-0.127	-0.112	0.096
							(-1.77)	(-1.97)	(-1.25)	(-1.22)	(-1.17)	(0.59)
internal stability								0.018	0.032	0.033	0.034	-0.012
								(0.48)	(0.80)	(0.79)	(0.80)	(-0.58)
external stability									-0.149*	-0.145*	-0.153*	-0.130**
									(-1.96)	(-1.99)	(-2.00)	(-2.31)
lack of ethnical conflict										-0.019	-0.017	-0.099
										(-0.25)	(-0.22)	(-1.14)
democracy											-0.024	-0.061
											(-0.38)	(-0.97)
Ave. military spending of neighbors												-1.109*
												(-1.86)
Obs.	174	174	174	174	174	174	174	174	174	174	174	174
R-sq	0.79	0.79	0.80	0.80	0.80	0.80	0.81	0.81	0.82	0.82	0.82	0.86

Robust *t-statistics* are in parentheses (clustered standard errors at country level). All independent variables are lagged by one year. ***, **, and * indicate significance at the 1%, 5%, and 10% levels, respectively. All variables (except for ICRG indexes) are in logarithmic form. Country and time-fixed effects are included in all models.

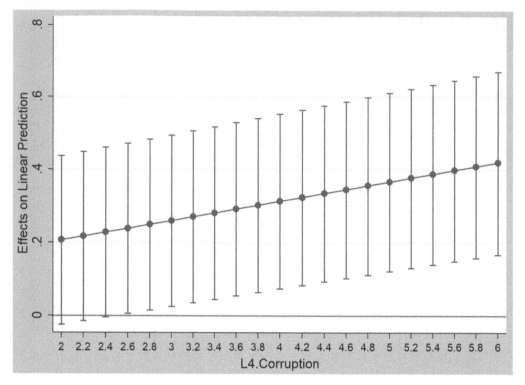

Figure 2: Marginal effect of oil rents on military spending per capita at different levels of corruption in non-GCC countries of the MENA.
Note: The marginal effects are based on Model 12 in Table 3. The solid middle line shows the marginal effects and dashed lines are confidence intervals at 90% level. L4 stands for a four-year lag of variable.

5 Conclusion

With the aim of understanding the connection between oil wealth and military budgets in the GCC and non-GCC countries of the MENA region, we investigated how the effect of increases in oil rents on military spending might be contingent on the level of corruption in this region. To test this hypothesis, we employed panel data covering the 1984–2014 period for five GCC and ten non-GCC countries in the MENA region. Our theoretical expectation was supported by the data.

The final effect of oil rents on military spending in both GCC and non-GCC samples in the MENA region depends on the level of political corruption. At higher levels of corruption, oil rents have a higher effect on military spending. In the GCC sample, the overall effect of increasing oil rents on military spending is negative. This is in line with observations of Bjorvatn and Farzanegan on the positive effects of oil rents on stability in countries with a sufficiently powerful incumbent. In the non-GCC sample of MENA, the overall effect of oil rents on military spending at different levels of corruption is positive.

Our main results hold when we control for a set of internal and external determinants of military spending, such as real income per capita, education, size of population, intensity of trade, share of total government spending in the economy, risk of internal and external conflicts, risk of ethnic tensions, rule of law, and quality of democratic institutions. Furthermore, our main results are not biased, due to omitted country- and year-specific factors. Our findings help to better understand the complex association between oil rents and militarization of economies in the GCC/MENA region. Corruption and its development over time matters significantly in this nexus.

ORCID

Mohammad Reza Farzanegan ⓘ http://orcid.org/0000-0002-6533-3645

Bibliography

1 Primary sources

Howell, Llewellyn D., "International Country Risk Guide Methodology", *The PRS Group* (2015), available online at www.prsgroup.com/wp-content/uploads/2012/11/icrgmethodology.pdf.

Organization of the Petroleum Exporting Countries (OPEC), "Annual Statistical Bulletin (ASB)" (2017), available online at http://asb.opec.org/.

World Bank, "The Changing Wealth of Nations: Measuring Sustainable Development in the New Millennium" (2011), available online at http://documents.worldbank.org/curated/en/6301814683396 56734/pdf/588470PUB0Weal101public10BOX353816B.pdf.

———, "World Development Indicators" (2017), available online at https://openknowledge.worldbank.org/handle/10986/26447.

2 Secondary sources

Abbas, Tehmina; Eva Anderson; Katherine Dixon; Emily Knowles; Gavin Raymond; and Leah Wawro, "Regional Results Middle East & North Africa: Government Defence Anti-Corruption Index 2015", *Transparency International Defence and Security* (2015), available online at http://government. defenceindex.org/downloads/docs/GI-MENA-Regional-Results-web.pdf.

Alesina, Alberto; Arnaud Devleeschauwer; William Easterly; Sergio Kurlat; and Romain Wacziarg, "Fractionalization", *Journal of Economic Growth* 8.2 (2003), pp. 155–94, available online at https://dash.harvard.edu/bitstream/handle/1/4553003/alesinassrn_fractionalization.pdf.

Andersen, Jørgen and Silje Aslaksen, "Oil and Political Survival", *Journal of Development Economics* 100.1 (2013), pp. 89–106, available online at www.sv.uio.no/econ/personer/vit/siljeasl/oil%20and%20political%20survival.pdf.

Asseery, Ahmed A. A., "Evidence from Time Series on Militarising the Economy: The Case of Iraq", *Applied Economics* 28.10 (1996), pp. 1257–61.

Atkinson, Giles and Kirk Hamilton, "Savings, Growth and the Resource Curse Hypothesis", *World Development* 31.11 (2003), pp. 1793–807.

Barro, Robert J., "Determinants of Democracy", *Journal of Political Economy* 107.6 (1999), pp. 158–83, available online at https://dash.harvard.edu/bitstream/handle/1/3451297/barro_determinantsdemocracy .pdf?sequence=2.

Besley, Timothy and Andrea Prat, "Handcuffs for the Grabbing Hand? Media Capture and Government Accountability", *American Economic Review* 96.3 (2006), pp. 720–36.

Bjorvatn, Kjetil and Mohammad Reza Farzanegan, "Demographic Transition in Resource Rich Countries: A Bonus or a Curse?", *World Development* 45 (2013), pp. 337–51.

———, "Resource Rents, Balance of Power, and Political Stability", *Journal of Peace Research* 52.6 (2015), pp. 758–73.

———, "Natural-Resource Rents and Political Stability in the Middle East and North Africa", *CESifo DICE Report* 13.3 (2015), pp. 33–7, available online at www.cesifo-group.de/DocDL/dice-report-2015-3_ Bjorvatn-Farzanegan_October.pdf.

Brunetti, Aymo and Beatrice Weder, "A Free Press Is Bad News for Corruption", *Journal of Public Economics* 87.7–8 (2003), pp. 1801–24.

Castello-Climent, Amparo, "On the Distribution of Education and Democracy", *Journal of Development Economics* 87.2 (2008), pp. 179–90.

Chan, Steve, "Defense Burden and Economic Growth: Unraveling the Taiwanese Enigma", *The American Political Science Review* 82.3 (1988), pp. 913–20.

Collier, Pail and Anke Hoeffler, "On Economic Causes of Civil War", *Oxford Economic Papers* 50.4 (1998), pp. 563–73, available online at https://asso-sherpa.org/sherpa-content/docs/programmes/GDH/Campagne_RC/War.pdf.

———, "Military Spending and the Risks of Coups d'Etat", Centre for the Study of African Economies, Oxford University (March 2007), available online at http://web.worldbank.org/archive/website01241/WEB/IMAGES/MILITARY.PDF.

Cruz, Cesi; Philip Keefer; and Carlos Scartascini, "Database of Political Institutions Codebook, 2015 Update (DPI2015)", *Inter-American Development Bank* (2016), available online at https://publications.iadb.org/handle/11319/7408.

Davoodi, Hamid; Benedict Clements; Jerald Schiff; and Peter Debaere, "Military Spending, the Peace Dividend, and Fiscal Adjustment", working paper submitted to the *International Monetary Fund* 99.87, Washington, DC (1999), available online at www.imf.org/external/pubs/ft/wp/1999/wp9987.pdf.

Dizaji, S. F.; Mohammad Reza Farzanegan; and Alireza Naghavi, "Political Institutions and Government Spending Behavior: Theory and Evidence from Iran", *International Tax and Public Finance* 23.3 (2016), pp. 522–49.

Donchev, Dilyan and Gergely Ujhelyi, "What Do Corruption Indices Measure?", *Economics and Politics* 26.2 (2014), pp. 309–331.

Dunne, J. Paul, and Sam Perlo-Freeman, "The Demand for Military Spending in Developing Countries: A Dynamic Panel Analysis", *Defence and Peace Economics* 14.6 (2003), pp. 461–74, available online at https://pdfs.semanticscholar.org/d6ad/94226ac0f6962c82d1c92089abeb65cd6469.pdf.

Dunne, J. Paul, and Sam Perlo-Freeman, "The Demand for Military Spending in Developing Countries", *International Review of Applied Economics* 17.1 (2003), pp. 23–48.

Dunne, J. Paul, Sam Perlo-Freeman, and Ron Smith P., "The Demand for Military Expenditure in Developing Countries: Hostility Versus Capability", *Defence and Peace Economics* 19.4 (2008), pp. 293–302.

Farzanegan, Mohammad Reza, "Illegal Trade in the Iranian Economy: Evidence from Structural Equation Model", *European Journal of Political Economy* 25.4 (2009), pp. 489–507.

———, "Oil Revenues Shocks and Government Spending Behavior in Iran", *Energy Economics* 33.6 (2011), pp. 1055–1069.

———, "Military Spending and Economic Growth: The Case of Iran", *Defence and Peace Economics* 25.3 (2014), pp. 247–69. doi:10.1080/10242694.2012.723160.

Farzanegan, Mohammad Reza; Christian Lessmann; and Gunther Markwardt, "Natural Resource Rents and Internal Conflicts: Can Decentralization Lift the Curse?", *Economic Systems* 42.2 (2018), pp. 186–205.

Farzanegan, Mohammad Reza and Stefan Witthuhn, "Corruption and Political Stability: Does the Youth Bulge Matter?", *European Journal of Political Economy* 49 (2017), pp. 47–70.

Fisman, Raymond and Edward Miguel, "Corruption, Norms, and Legal Enforcement: Evidence from Diplomatic Parking Tickets", *Journal of Political Economy* 115.6 (2007), pp. 1020–48.

Fisman, Raymond and Shang-Jin Wei, "The Smuggling of Art, and the Art of Smuggling: Uncovering the Illicit Trade in Cultural Property and Antiques", *American Economic Journal: Applied Economics* 1.3 (2009), pp. 82–96, available online at https://pdfs.semanticscholar.org/3a14/2d3105f32130e30ac5d890f87c011b1554b3.pdf.

Fjelde, Hanne, "Buying Peace? Oil Wealth, Corruption and Civil War, 1985–99", *Journal of Peace Research* 46.2 (2009), pp. 199–218.

Fjelde, Hanne and Håvard Hegre, "Political Corruption and Institutional Stability", *Studies in Comparative International Development* 49.3 (2014), pp. 267–99, available online at http://folk.uio.no/hahegre/Papers/DemDep_10032011.pdf.

Glaeser, Edward; Giacomo Ponzetto; and Andrei Shleifer, "Why Does Democracy Need Education?", working paper submitted to the *National Bureau of Economic Research* 12128, Cambridge (2006), available online at https://scholar.harvard.edu/files/glaeser/files/democracy_final_jeg_1.pdf.

Gupta, Sanjeev; Luiz de Mello; and Raju Sharan, "Corruption and Military Spending", *European Journal of Political Economy* 17.4 (2001), pp. 749–77, available online at www.imf.org/external/pubs/ft/wp/2000/wp0023.pdf.

Hessami, Zohal, "Political Corruption, Public Procurement, and Budget Composition: Theory and Evidence from OECD Countries", *European Journal of Political Economy* 34 (2014), pp. 372–89.

Hewitt, Daniel, "Military Expenditures Worldwide: Determinants and Trends, 1972–88", *Journal of Public Policy* 12.2 (1992), pp. 105–52.

Lebovic, James H., and Ashfaq Ishaq, "Military Burden, Security Needs and Economic Growth in the Middle East", *Journal of Conflict Resolution* 31.1 (1987), pp. 106–38.

Majeski, Stephen J., and David L. Jones, "Arms Race Modeling: Causality Analysis and Model Specification", *Journal of Conflict Resolution* 25.2 (1981), pp. 259–88.

Manzetti, Luigi and Carole J. Wilson, "Why Do Corrupt Governments Maintain Public Support", *Comparative Political Studies* 40.8 (2007), pp. 949–970.

Martin, Philippe; Thierry Mayer; Mathias Thoenig, "Does Globalisation Pacify International Relations?", *Vox* (2007), available online at http://voxeu.org/article/trade-andor-war.

Mauro, Paolo, "Corruption and the Composition of Government Expenditure", *Journal of Public Economics* 69.2 (1998), pp. 263–79, available online at http://darp.lse.ac.uk/PapersDB/Mauro_(JPubE_98).pdf.

Mintz, Alex and Chi Huang, "Defense Expenditures, Economic Growth and the 'Peace Dividend'", *American Political Science Review* 84.4 (1990), pp. 1283–93.

Morris, Loveday, "Investigation Finds 50,000 'Ghost' Soldiers in Iraqi Army, Prime Minister Says", *The Washington Post*, 30 November 2014, available online at www.washingtonpost.com/world/middle_east/investigation-finds-50000-ghost-soldiers-in-iraqi-army-prime-minister-says/2014/11/30/d8864d6c-78ab-11e4-9721-80b3d95a28a9_story.html?noredirect = on&utm_term = .24c4f43b0c19.

Olken, Benjamin A., "Corruption Perceptions vs. Corruption Reality", *Journal of Public Economics* 93.7–8 (2009), pp. 950–64, available online at https://economics.mit.edu/files/3931.

Rose-Ackerman, Susan, *Corruption and Government: Causes, Consequences and Reform* (Cambridge: Cambridge University Press, 1999).

Rosh, Robert M., "Third World Militarization: Security Webs and the States They Ensnare", *Journal of Conflict Resolution* 32.4 (1988), pp. 671–98.

Sachs, Jeffrey D., and Andrew M. Warner, "The Curse of Natural Resources", *European Economic Review* 45.4–6 (2001), pp. 827–38.

Shleifer, Andrei and Robert W. Vishny, "Corruption", *Quarterly Journal of Economics* 108.3 (1993), pp. 599–617, available online at https://projects.iq.harvard.edu/gov2126/files/shleifer_and_vishy.pdf.

Sung, Hung-En, "A Convergence Approach to the Analysis of Political Corruption: A Cross-National Study", *Crime, Law and Social Change* 38.2 (2002), pp. 137–60, available online at www.researchgate.net/profile/Hung-En_Sung/publication/226565521_A_convergence_approach_to_the_analysis_of_political_corruption_A_cross-national_study/links/0fcfd50bc9ba63fd88000000/A-convergence-approach-to-the-analysis-of-political-corruption-A-cross-national-study.pdf .

Tanzi, Vito, "Corruption Around the World: Causes, Consequences, Scope, and Cures", *IMF Staff Papers* 45.4 (1998), pp. 559–94, available online at http://syahia.atreides.online.fr/iae/ethics/tanzi.pdf.

Treisman, Daniel, "The Causes of Corruption: A Cross-National Study", *Journal of Public Economics* 76.3 (2000), pp. 399–457, available online at http://citeseerx.ist.psu.edu/viewdoc/download?doi = 10.1.1.8.4980&rep = rep1&type = pdf.

Wooldridge, Jeffrey M., *Econometric Analysis of Cross Section and Panel Data* (Cambridge, MA: MIT Press, 2002).

Yildirim, Jülide; Selami Sezgin; and Nadir Öcal, "Military Expenditure and Economic Growth in Middle Eastern Countries: A Dynamic Panel Data Analysis", *Defence and Peace Economics* 16.4 (2005), pp. 283–95.

Index